HOLT
Chemistry
The Physical Setting

Focus on the Regents Exam
Review Guide with Practice Exams

International Arts Business School
@ Wingate Campus
600 Kingston Avenue
Brooklyn, NY 11203

HOLT, RINEHART AND WINSTON
A Harcourt Education Company

Orlando • **Austin** • New York • San Diego • Toronto • London

Copyright © by Holt, Rinehart and Winston.

All rights reserved. No part of this publication may be reproduced or transmitted in any form or by any means, electronic or mechanical, including photocopy, recording, or any information storage and retrieval system, without permission in writing from the publisher.

Requests for permission to make copies of any part of the work should be mailed to the following address: Permissions Department, Holt, Rinehart and Winston, 10801 N. MoPac Expressway, Building 3, Austin, Texas 78759.

Printed in the United States of America

ISBN 0-03-036206-7

1 2 3 4 5 6 7 862 09 08 07 06 05 04

Contents

Unit I	1
Unit II	17
Unit III	29
Unit IV	41
Unit V	55
Unit VI	69
Unit VII	85
Unit VIII	103
Unit IX	115
Unit X	123
Unit XI	135
Unit XII	145
Unit XIII	157
Unit XIV	171
Unit XV	185
Measurement and Calculations	199
Reference Tables	207
Past Regents Exams	228
Glossary	281
Index	287

Name _____ Class _____ Date _____

UNIT I

Matter and Energy

Holt Chemistry: The Physical Setting

Chemistry is the study of matter. Chemists investigate both the properties of matter and the changes that matter undergoes. Chemistry, however, is not limited to a laboratory. In fact, chemistry is continuously happening all around you. Everything around you is made of matter. Even you are made of matter.

States of Matter

Matter is the stuff or material that makes up you and everything around you, including this book and the chair you may be sitting on. Scientists define **matter** as anything that has **mass** and **volume**. Mass is the quantity of matter in an object or substance. Volume is the space that the mass takes up.

Mass is not the same as **weight**. The weight of an object is defined as the force produced by gravity acting on an object. The mass of an object remains constant no matter where the object is located. However, the weight of the object does depend on its location because gravity can vary from one location to another. Although weight and mass are not equivalent, the weight of an object can tell you something about its mass. If you double the mass of an object and keep it in the same place, the weight of the object will also be doubled.

All matter is made of particles. The type and arrangement of these particles in a sample of matter determines the properties of that matter. These particles are arranged in different ways to form the **states of matter**. The states of matter are the various physical forms of matter. These include solid, liquid, and gas.

Solids have a definite shape and a definite volume. Solids have these properties because the particles that make up a solid are held tightly in a rigid structure. The particles vibrate, but they do not move from their fixed positions.

Liquids have a definite volume but not a definite shape. The particles in a liquid are not held together as tightly as those in a solid. Rather, the particles in a liquid can slip past one another. As a result, a liquid flows to take the shape of its container.

Gases have neither a definite shape nor a definite volume. Gases have these properties because their particles only weakly attract one another. Instead, the particles in a gas move independently at high speed. As a result, a gas takes the shape and volume of any container it occupies.

What You'll Need to Learn

This topic is part of the Regents Curriculum for the Physical Setting Exam.
Standard 4, Performance Indicator 3.1: *Explain the properties of materials in terms of the arrangement and properties of the atoms that compose them.*
Standard 4, Performance Indicator 4.1: *Observe and describe transmission of various forms of energy.*
Standard 4, Performance Indicator 4.2: *Explain heat in terms of kinetic molecular theory.*

Which Terms You'll Need to Know

atom
chemical change
chemical property
chemical reaction
compound
density
element
endothermic
exothermic
energy
heat
heterogeneous
homogeneous
mass
matter
mixture
physical change
physical property

Copyright © by Holt, Rinehart and Winston. All rights reserved.

Name _____ Class _____ Date _____

Unit I Matter and Energy *continued*

Which Terms You'll Need to Know
product
reactant
states of matter
temperature
volume
weight

Where You Can Learn Even More

Holt Chemistry: The Physical Setting
Chapter 1: The Science of Chemistry
Chapter 2: Matter and Energy

Models can be used to show how the particles are arranged differently in a solid, liquid, and gas. For example, observe how particles of water are arranged differently in ice (solid), water (liquid), and vapor (gas).

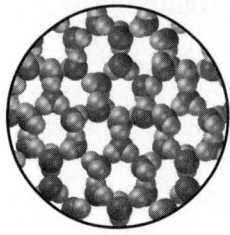
solid

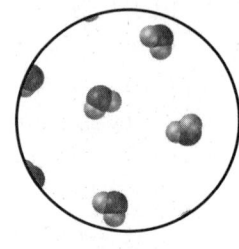
liquid   gas

REVIEW YOUR UNDERSTANDING

_____ 1. Which is the state of matter where the particles take the shape and volume of the container?
  (1) solid
  (2) liquid
  (3) gas
  (4) liquid and gas, both

_____ 2. What does all matter possess?
  (1) mass and volume
  (2) particles that move freely
  (3) mass, but not volume
  (4) volume, but not mass

_____ 3. Identify the state of matter where the particles can only vibrate in their fixed positions.
  (1) solid
  (2) liquid
  (3) gas
  (4) solid and liquid, both

_____ 4. Which state of matter is represented in the following particle model?

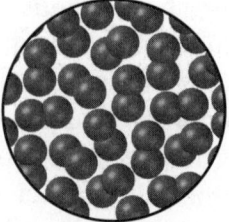

  (1) solid
  (2) liquid
  (3) gas
  (4) either a solid or liquid

Unit I Matter and Energy continued

Physical Properties and Physical Changes

States of matter are examples of **physical properties**. A physical property is a characteristic that can be determined without changing the nature of the substance. For example, you can look at a piece of chalk and determine that it is a solid. You can also tell that it is white, hard, and smooth. Therefore, color, hardness, and smoothness are also physical properties.

Density is another physical property of matter. Density is the ratio of mass to volume.

$$density = mass/volume \text{ or } D = m/V$$

The density of a substance is the same no matter what the size of the sample is. Knowing the density of a sample can help you identify what it is. For example, if you calculate the density of a shiny, yellow-colored metal as $8.91 g/cm^3$, you can reasonably conclude that the metal is copper, whose density is $8.94 \ g/cm^3$. The metal is certainly not gold, whose density is $19.3 \ g/cm^3$.

Self-Check Explain why density is a physical property of matter.

You can change the physical properties of a sample of matter. For example, if you crush a piece of chalk with a hammer, it will no longer be hard and smooth. Crushing the chalk causes a **physical change**. The physical properties of the chalk have changed, but it is still chalk. A physical change does not involve a change in a substance's identity.

Changes in the state of matter always involve physical changes. Consider what happens when an ice cube melts. As the ice cube changes state, its physical properties also change. The matter is no longer solid and hard, but liquid and wet.

Energy is required to bring about a change in state. Energy is defined as the capacity to do work, such as moving an object or generating light. Every change in matter involves a change in energy. Some changes in matter absorb energy. Such changes are known as **endothermic** processes.

The melting of ice and the boiling of water are two examples of endothermic changes in matter. Energy must be absorbed by ice to melt and by water to boil.

Some changes in matter release energy. Such changes are known as **exothermic** processes. The condensation of water vapor and the freezing of water are two examples of exothermic changes in matter. Energy is released by vapor as it liquefies and by water as it freezes.

Notes/Study Ideas/Answers

Unit I Matter and Energy *continued*

Self-Check What is this difference between an endothermic process and an exothermic process?

Energy can be absorbed and released in different forms. These forms include chemical, mechanical, light, electrical, sound, and heat. **Heat** is defined as the energy transferred between objects that are at different **temperatures**. Heat energy is always transferred from a warmer object to a cooler object.

Heat is different from temperature. Heat is a transfer of energy, usually in a form called thermal energy. Thermal energy is associated with the random motion of the particles in an object. Energy association with motion is called *kinetic energy*. Temperature is a measurement of the random motion of the particles in a sample of matter. The more rapidly the particles move, the greater their kinetic energy is. The greater their kinetic energy, the higher the temperature is. Conversely, a lower temperature indicates that the particles have less kinetic energy.

Self-Check Explain the difference between heat and temperature.

The transfer of energy as heat can result in a higher temperature. For example, if you place a thermometer in a beaker of water and heat it, you will observe an increase in temperature. However, the transfer of energy as heat does not always result in a higher temperature. Imagine that you place some ice cubes and water in a sealed, insulated container that is slowly heated. Even though heat is being transferred to the ice/water mixture, you would not notice any increase in temperature. Once all the ice has melted, however, the temperature will start to increase.

A graph can be drawn to show the relationship between temperature and the energy being added as heat to bring about a change in state. Examine the graph below that represents a heating curve for water.

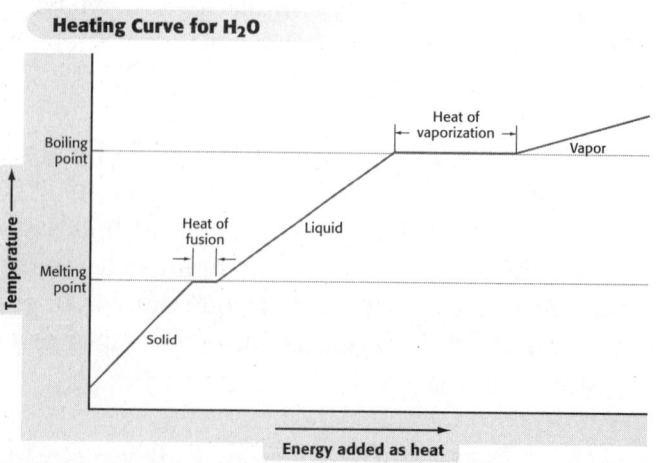

Unit I Matter and Energy continued

Notice that the temperature of the ice (solid) increases as energy is being added as heat. However, the temperature does not increase as the ice melts, or undergoes a change in state. The energy being added as heat is being used to move the water particles farther apart from one another as the ice turns to liquid. The energy being added is called the *heat of fusion*. Fusion is another word for melting. Notice that the heat of fusion corresponds to the melting point of the solid. The melting point of a solid is also the freezing point of its liquid.

Self-Check Why is the heat of fusion drawn as a flat line on the graph?

Once all the ice has melted, the energy being added as heat raises the average kinetic energy of the water particles. This is measured as an increase in the temperature of the liquid.

As the liquid turns into a vapor, again there is no increase in temperature. The energy being added as heat at this point is being used to move the water particles farther apart from one another as the liquid turns into a vapor. This value on the graph is called the *heat of vaporization*. Notice that the heat of vaporization corresponds to the boiling point of a liquid. The boiling point of a liquid is also the condensation point of its vapor. Also notice that much more energy must be added as heat to vaporize water than to melt ice.

Self-Check Why is the flat line representing the heat of vaporization longer than the flat line representing the heat of fusion?

Once all the liquid has vaporized, the energy being added as heat raises the average kinetic energy of the vapor particles. This is measured as an increase in the temperature of the vapor as shown on the graph.

REVIEW YOUR UNDERSTANDING

_____ 5. Thermal energy is associated with
 (1) the random movement of particles.
 (2) exothermic processes, only.
 (3) endothermic processes, only.
 (4) gases, only.

_____ 6. If a process is exothermic, then energy
 (1) cannot be transferred as heat.
 (2) is absorbed.
 (3) is released.
 (4) is either absorbed or released, depending on the change in state.

Unit 1 Matter and Energy *continued*

_____ **7.** The term fusion describes
  (1) boiling and condensing.
  (2) condensing and melting.
  (3) melting and boiling.
  (4) freezing and melting.

_____ **8.** Which change in state is associated with the heat of vaporization?
  (1) solid to liquid
  (2) liquid to vapor
  (3) solid to vapor
  (4) liquid to solid

_____ **9.** Energy as heat is added while a sample of matter changes state. The temperature
  (1) increases.
  (2) decreases.
  (3) remains the same.
  (4) does not measure the average kinetic energy of the particles.

Chemical Properties and Chemical Changes

A **chemical property** of a sample of matter cannot be identified unless a change in the identity of the substance takes place. For example, oxygen is a colorless, odorless gas. These are physical properties of oxygen because you can identify them without changing oxygen in any way. Oxygen also supports combustion. This is an example of a chemical property because the only way to observe this physical property is to burn something in oxygen. When this is done, oxygen's identity is no longer the same.

When combustion occurs, oxygen undergoes a **chemical change.** A chemical change involves the formation of new substances. New substances are formed when a **chemical reaction** occurs. A chemical reaction is the process by which one or more substances change to produce one or more different substances. A chemical reaction always involves chemical changes.

Self-Check Explain why dissolving sugar in iced tea is not an example of a chemical change.

Chemical reactions happen all around you. A cake baking in the oven, gasoline burning in a car's engine, and turning on a flashlight all involve chemical reactions. Chemical reactions also happen inside you. Respiration, digestion, and circulation all depend on chemical reactions.

Unit I Matter and Energy *continued*

Chemical equations are used to show what happens during a chemical reaction. For example, the burning of gasoline in a car's engine can be described by the following word equation.

$$\text{octane} + \text{oxygen} \rightarrow \text{carbon dioxide} + \text{water}$$

In an equation, the substances on the left-hand side of the arrow are the **reactants**. They are the substances whose identity changes. Substances on the right-hand side of the arrow are the **products**. They are the new substances that are made in a chemical reaction.

> **Self-Check** Identify the reactants and products when gasoline burns in a car's engine.

Recall that changes in energy take place whenever a physical change occurs. Not surprisingly, changes in energy also take place when a chemical change occurs. Again, a chemical change may release energy, making the process exothermic. On the other hand, a chemical change may absorb energy, making the process endothermic. Chemical equations can be written to show if the reaction is exothermic or endothermic. For example, the burning of gasoline is an exothermic process as shown by the following equation.

$$\text{octane} + \text{oxygen} \rightarrow \text{carbon dioxide} + \text{water} + \text{energy}$$

Notice that energy is written on the right-hand side of the arrow, just as if it were a product. Writing energy as if it were a product indicates that the above reaction is exothermic.

> **Self-Check** Water can be broken down into hydrogen and oxygen. The process is endothermic. Write the word equation for this chemical reaction. Be asure to include the word energy in your equation.

A chemical reaction is a rearrangement of atoms. The products contain the same number and types of atoms as were found in the reactants. Atoms are never created or destroyed during a chemical reaction.

Notes/Study Ideas/Answers

Unit I Matter and Energy continued

REVIEW YOUR UNDERSTANDING

_____ 10. A chemical reaction
   (1) does not involve changes in energy.
   (2) involves a rearrangement of atoms to form new substances.
   (3) always releases energy, usually as either heat or light.
   (4) cannot be an endothermic process.

_____ 11. Any change in matter in which its identity is changed is known as a(n)
   (1) physical change.
   (2) chemical change.
   (3) endothermic process.
   (4) exothermic process.

_____ 12. The products of a chemical reaction
   (1) contain atoms that are not found in the reactants.
   (2) are shown on the left-hand side of the equation.
   (3) are shown on the right-hand side of the equation.
   (4) have the same arrangement of atoms as found in the reactants.

_____ 13. Examine the following word equation.

   sodium + chlorine → sodium chloride + energy

   Which statement is correct concerning this equation?
   (1) Sodium chloride is a reactant.
   (2) Sodium and chlorine are the new substances that are formed during this reaction.
   (3) This reaction is an endothermic process.
   (4) This reaction is an exothermic process.

_____ 14. If a chemical reaction is endothermic, then
   (1) energy is absorbed by the reaction.
   (2) energy is released by the reaction.
   (3 a reactant must undergo a change in state.
   (4) a product has the same chemical identity as the reactant.

Unit I Matter and Energy continued

Elements, Compounds, and Mixtures

Matter exists in so many different forms. Classifying matter into groups that share certain characteristics makes it easier to study it. One way to classify matter is by its state. A sample of matter can be classified as a solid, liquid, or gas.

There are other ways to classify matter. Another way is to classify matter as an element, compound, or mixture. The difference between these three classifications of matter can be either a physical or chemical property.

An **element** is a substance that cannot be broken down or separated into simpler substances by chemical means. An element contains only one kind of atom. An **atom** is the smallest unit of an element that maintains the properties of that element.

An element may consist of a single atom. An element may also consist of two or millions of atoms. However, if an element consists of more than one atom, all the atoms in that element are the same type. For example, the oxygen gas we breathe is an element that consists of two oxygen atoms. The ozone gas that protects us from the harmful effects of sunlight is an element that consists of three oxygen atoms.

A particle model can be used to illustrate an element. Notice that only one kind of atom is present in an element.

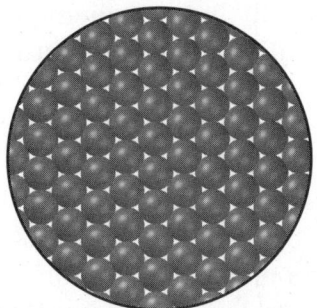

Copper atoms

Self-Check Carbon dioxide consists of carbon atoms and oxygen atoms. Why is carbon dioxide not classified as an element?

A **compound** is a substance made up of two or more different elements that are chemically joined. Unlike elements, compounds are composed of more than one kind of atom. However, compounds do have something in common with elements. Both elements and compounds are *pure substances*.

A pure substance has definite chemical and physical properties throughout a given sample, and from sample to sample. For example, the element gold always melts at 1064°C, and the compound water always consists of two hydrogen atoms and one oxygen atom that are joined together.

Unit I Matter and Energy continued

A **mixture** is a combination of two or more substances that are not chemically joined to one another. Air is a mixture of various gases, including nitrogen and oxygen. Brass is a mixture of the elements copper and zinc.

> **Self-Check** What do compounds and mixture have in common that make them both different from an element?

Particle models can be used to illustrate compounds and mixtures. Notice that both compounds and mixtures contain more than one kind of atom.

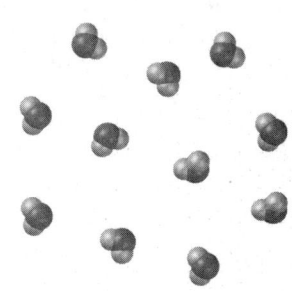

A compound is composed of two or more elements that are chemically joined. A mixture is composed of two or more substances that are physically but not chemically joined. As a result, there are two main differences between a compound and a mixture.

First, the properties of a compound are often very different from the properties of the elements that make them up. For example, the properties of water are very different from the elements, hydrogen and oxygen, that make up water.

In contrast, the properties of a mixture reflect the properties of the substances it contains. For example, the oxygen gas that is a component of the mixture called air can still support a candle flame.

Second, the elements that make up a compound always have a definite composition. For example, the ratio of hydrogen atoms to oxygen atoms in water is always 2:1.

In contrast, the substances that make up a mixture can be present in varying proportions. For example, the air in certain places may contain more oxygen than the air in other locations.

> **Self-Check** Does adding sand to water make a compound or mixture? Explain.

Unit I Matter and Energy continued

There are two types of mixtures. One type of mixture is called a **homogeneous mixture**. In a homogeneous mixture, the substances are distributed uniformly throughout the mixture. An example of a homogeneous mixture is sweetened iced tea. Stirring produces a homogeneous mixture by dissolving the sugar so that it is uniformly distributed throughout the ice tea.

What would happen if you continued to add sugar? At some point, no more sugar would dissolve, no matter how long you stirred the iced tea. The sugar that does not dissolve will slowly settle to the bottom of the glass. In this case, the sweetened iced tea represents a **heterogeneous mixture**. In a heterogeneous mixture, the substances are not uniformly distributed throughout the mixture.

Self-Check Explain how sand and water can make either a homogeneous mixture or a heterogeneous mixture.

No matter whether the mixture is homogeneous or heterogeneous, the components in a mixture can be separated based on their physical properties. For example, you can use filtration to separate the components based on their size. You can also use distillation and evaporation, which depend on differences in boiling points.

REVIEW YOUR UNDERSTANDING

_____ 15. Which contains only one type of atom?
(1) element
(2) compound
(3) homogeneous mixture
(4) heterogeneous mixture

_____ 16. All compounds
(1) contain one element, only.
(2) can contain no more than two elements.
(3) can be broken down into simpler substances by chemical means.
(4) can be broken down into simpler substances by physical means.

_____ 17. Which substance cannot be decomposed or separated into simpler substances?
(1) oxygen
(2) water
(3) sweetened ice tea
(4) air

Notes/Study Ideas/Answers

Unit I Matter and Energy *continued*

Notes/Study Ideas/Answers

?ED/#s 18 & 19 will not fit on the previous page

_____ 18. The solid and liquid components in a mixture are separated from one another based on
(1) their chemical properties.
(2) their physical properties.
(3) the chemical composition.
(4) the types of atoms they contain.

_____ 19. Which is *not* a pure substance?
(1) oxygen
(2) carbon dioxide
(3) air
(4) nitrogen

ANSWERS TO SELF-CHECK QUESTIONS

- Density is a physical property because it can be determined without changing the nature of the substance.

- An endothermic process absorbs energy. An exothermic process releases energy.

- Heat is the energy that is transferred between objects that are at different temperatures. Temperature is a measurement of the average kinetic energy of the particles in a sample of matter.

- Heat of fusion is drawn as a flat line because there is no change in temperature even though energy is being added as heat.

- The heat of vaporization flat line is longer because more energy must be added to change water into a vapor than to change ice into a liquid (heat of fusion).

- Dissolving sugar in iced tea is not an example of a chemical change because no new substances are formed.

- Octane and oxygen are the reactants. Carbon dioxide and water are the products.

- water + energy → hydrogen + oxygen

- Carbon dioxide consists of two different kinds of atoms and can be broken down by chemical means into simpler substances.

- Both compounds and mixtures contain two or more different kinds of atoms.

Unit I Matter and Energy *continued*

- Adding sand to water produces a mixture. Both the sand and water retain their properties, and they can be mixed in any proportion.

- If the sand and water are continuously stirred, then you can have a homogeneous mixture. Once the sand begins to settle, then you have a heterogeneous mixture.

Notes/Study Ideas/Answers

| UNIT I | Holt Chemistry: The Physical Setting |

Questions for Regents Practice

Matter and Energy

PART A

_____ 1. Identify the state of matter where the particles are closest to one another.
(1) solid
(2) liquid
(3) gas
(4) vapor

_____ 2. Which has a definite volume but not a definite shape?
(1) solid
(2) liquid
(3) gas
(4) both a liquid and a gas

_____ 3. A physical change
(1) involves a change in the identity of a substance.
(2) occurs when matter changes state.
(3) occurs when a compound is decomposed into simpler substances.
(4) does not involve changes in energy.

_____ 4. The heat of fusion represents the quantity of energy that is
(1) released when water boils.
(2) required to change ice into water.
(3) absorbed when water vapor condenses into a liquid.
(4) required to join elements to form a compound.

_____ 5. Which is *not* a form of energy?
(1) light
(2) mechanical
(3) heat
(4) temperature

_____ 6. Which change is exothermic?
(1) melting of iron
(2) vaporization of alcohol
(3) freezing of water
(4) turning a solid directly into a gas

_____ 7. Which term refers to matter that can be heterogeneous?
(1) element
(2) compound
(3) mixture
(4) atom

_____ 8. Which is a characteristic of all compounds?
(1) They consist of elements whose proportions can vary.
(2) Their chemical composition consists of atoms present in a definite ratio.
(3) They are heterogeneous.
(4) They cannot be broken down into simpler substances.

_____ 9. Separating the components in a mixture
(1) requires energy when bonds are broken.
(2) releases energy when bonds are broken.
(3) depends on their chemical properties.
(4) depends on their physical properties.

Copyright © by Holt, Rinehart and Winston. All rights reserved.

Matter and Energy

Unit I Matter and Energy continued

PART B-1

_____ 10. Which state of matter does the following particle model represent?

(1) solid
(2) liquid
(3) gas
(4) heterogeneous mixture

_____ 11. Examine the graph below which represents the changes of state for an unknown substance.

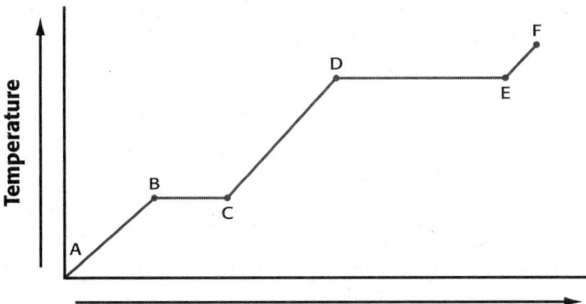

Which line segment represents the heat of fusion?
(1) AB
(2) BC
(3) CD
(4) DE

_____ 12. Heat differs from temperature in that heat
(1) can be transferred between objects as a form of energy.
(2) is a measurement of the average kinetic energy of the particles in a sample of matter.
(3) can be released only as a result of a chemical change.
(4) can be absorbed only as a result of a physical change.

_____ 13. Which is an example of a chemical change?
(1) evaporating alcohol
(2) melting wax
(3) burning wood
(4) condensing water

_____ 14. Identify the product(s) in the following chemical reaction.

hydrogen + oxygen → water

(1) water
(2) oxygen
(3) water and oxygen
(4) hydrogen and oxygen

_____ 15. Which contains substances that can be separated by filtration?
(1) homogeneous mixtures, only
(2) heterogeneous mixtures, only
(3) both homogeneous and heterogeneous mixtures
(4) the elements present in a compound

PART B-2

16. Why must energy be supplied to change a liquid into a vapor?

_____

_____

_____

Unit I Matter and Energy continued

17. Draw two models that illustrate the difference between a compound and a mixture. Both the compound and mixture contain two types of particles, A and B. Use the following symbols to represent the particles.
particle A: ●   particle B: ○

18. Draw two models that illustrate the difference between a homogenous mixture and a heterogeneous mixture. Both mixtures contain two components, A and B. Use the following symbols to represent the components.
particle A: ●   particle B: ○

PART C

19. When camping with your family, you boil a pot of water over a campfire. Describe the physical and chemical changes that take place during this process.

20. A student leaves an uncapped watercolor marker on an open notebook. Later, the student discovers that the leaking marker has produced a rainbow of colors. Should the ink be classified as an element, compound or mixture? Explain your answer.

Name _____ Class _____ Date _____

UNIT II

Atomic Structure

Holt Chemistry: The Physical Setting

An **atom** is the smallest unit of an element that retains the properties of that element. The ancient Greeks first suggested that all matter is made of atoms. However, they had no experimental proof to support their position. John Dalton, an English schoolteacher, again stated the belief that all matter is made of atoms in 1808.

Unlike the ancient Greeks, however, Dalton used experimental evidence that others had gathered to support his statement. This evidence came from experiments that investigated the chemical composition of substances and how substances interacted during a chemical reaction. Dalton developed his atomic theory based mainly on the results of these experiments. His theory explained most of the chemical data that existed during his time. However, later experiments would reveal that Dalton's atomic theory was not entirely correct.

Parts of an Atom

Dalton believed that an atom could not be broken down into smaller parts. Experiments performed in the mid-1800s, however, revealed that atoms could be broken down into pieces. These smaller parts are known as subatomic particles. There are three major subatomic particles.

One of these subatomic particles is a **proton**. Experiments revealed that a proton has a positive charge. The charge of a proton is usually written as +1.

Another subatomic particle is a **neutron**. A neutron was found to be a neutral particle. The charge of a neutron is usually written as 0.

The protons and neutrons are located in the central region of an atom. This region of an atom is called the **nucleus**. The nucleus is the dense, central portion of an atom where all its protons and neutrons are found. A nucleus then has an overall positive charge because of the protons it contains.

An atom, however, is neutral. This means that an atom must contain a subatomic particle that has a negative charge. The negative charges on these particles would balance the positive charges of the protons in the nucleus, making an atom neutral. The subatomic particle that has a negative charge is the **electron**. The charge of an electron is usually written as −1. The electron was actually the first subatomic particle to be discovered.

What You'll Need to Learn

This topic is part of the Regents Curriculum for the Physical Setting Exam.
Standard 4, Performance Indicator 3.1: *Explain the properties of materials in terms of the arrangement and properties of the atoms that compose them.*

Which Terms You'll Need to Know

atom
bright-line spectrum
electron
excited state
ground state
neutron
nucleus
orbital
proton

Where You Can Learn Even More:

Holt Chemistry: The Physical Setting
Chapter 3: Atoms and Moles

Copyright © by Holt, Rinehart and Winston. All rights reserved.

Unit II Atomic Structure continued

Properties of an Electron

Name	Symbol	As shown in figures	Charge	Common charge notation	Mass (kg)
Electron	e, e^-, or $_{-1}^{0}e$		-1.602×10^{-19} C	-1	9.109×10^{-31} kg

Properties of a Proton and a Neutron

Name	Symbol	As shown in figures	Charge	Common charge notation	Mass (kg)
Proton	p, p^+, or $_{+1}^{1}p$		$+1.602 \times 10^{-19}$ C	$+1$	1.673×10^{-27} kg
Neutron	n or $_{+0}^{1}n$		0 C	0	1.675×10^{-27} kg

The number of electrons is equal to the number of protons in an atom. Therefore, if you know how many protons an atom contains, then you also know how many electrons it has, and vice versa. For example, an atom that has 17 protons must also have 17 electrons. However, you cannot determine the number of neutrons an atom has even if you know how many protons or electrons it contains.

> **Self-Check** An atom contains 46 electrons. How many protons does it contain? Can you determine how many neutrons this atom has?

Soon after the subatomic particles were discovered, scientists began to build models of the atom. One of the first models proposed that an atom consisted of negatively-charged electrons embedded in a ball of positively-charged protons. However, this model was discarded when scientists obtained experimental results that pointed to a very different arrangement of the subatomic particles.

This newer model proposed that the electrons are not located in the nucleus. Rather, the electrons orbit the nucleus of an atom. The model also suggested that the electrons take up most of the space that an atom occupies. In contrast, the nucleus is only a very small fraction of the volume of an atom. Even though an atom has a volume, it is mostly empty space.

Unit II Atomic Structure continued

REVIEW YOUR UNDERSTANDING

_____ 1. Which subatomic particles are located in the nucleus?
 (1) electrons and protons
 (2) electrons and neutrons
 (3) protons and neutrons
 (4) electrons, only

_____ 2. If an atom contains 31 protons, then it
 (1) must be positively charged.
 (2) also contains 31 neutrons.
 (3) cannot be neutral.
 (4) has an equal number of electrons.

_____ 3. Which accounts for most of the volume of an atom?
 (1) electrons
 (2) nucleus
 (3) protons
 (4) protons and neutrons

_____ 4. Identify the subatomic particle whose charge is written as +1.
 (1) a neutron
 (2) a proton
 (3) an electron
 (4) either an electron or a proton

Energy Levels, Orbitals, and Shells

Scientists proposed a model of the atom where the electrons revolve around the nucleus in circular or elliptical orbits, much like the planets orbit the sun. However, this model could not explain why a negatively-charged electron did not get pulled into the nucleus by the positively-charge protons. In other words, this model could not explain why an atom did not collapse.

In 1913, a Danish physicist named Niels Bohr proposed another model of the atom. Rather than describe electrons in terms of orbits, Bohr described electrons in terms of their energy levels. His model proposed that electrons could be only certain distances from the nucleus. Each distance corresponds to a certain quantity of energy that an electron can have.

An electron that is as close to the nucleus as it can be is in its lowest energy level. The farther the electron is from the nucleus, the higher its energy level. Bohr's model states that an electron can be located in only one energy level or another. The electron can never be between energy levels.

Unit II Atomic Structure *continued*

Notes/Study Ideas/Answers

Bohr's model of the atom had to be modified when scientists discovered that electrons act like both particles and waves. This discovery led to the current model of the atom. This model depicts electrons as being in **orbitals**. An orbital is a region around the nucleus that corresponds to a specific energy level. An orbital is a region where an electron is likely to be found. It has no sharp boundaries. An orbital with a specific energy level is also referred to as an electron shell.

> **Self-Check** Compare the distance that an electron is from the nucleus to the quantity of energy that it has.

Electrons do not remain in just one shell. Rather they constantly move from one shell to another. Energy must be supplied to move an electron to a shell that is farther from the nucleus. An electron can also move to a shell that is closer to the nucleus. When an electron moves to a shell closer to the nucleus, energy is released.

Each electron in an atom has its own distinct quantity of energy. An electron that is in the lowest possible energy level, or in a shell closest to the nucleus that is available, is in the **ground state**. If an electron is not in its lowest possible energy level, then the electron is in an **excited state**. In an excited state, an electron is in a shell farther from the nucleus even though one closer to the nucleus is available.

REVIEW YOUR UNDERSTANDING

_____ **5.** Bohr described electrons in terms of their
  (1) charge.
  (2) energy level.
  (3) volume.
  (4) distance from the nucleus.

_____ **6.** Electrons orbit the nucleus in
  (1) specific energy levels known as shells.
  (2) paths much like those that the planets follow as they orbit the sun.
  (3) shells that are as far away as possible from the nucleus.
  (4) shells that always have the highest energy levels possible.

Unit II Atomic Structure continued

_____ **7.** An electron in the ground state
(1) is as far from the nucleus as possible.
(2) can move to a lower energy level.
(3) is in the closest available shell to the nucleus.
(4) releases energy when it moves to a higher energy level.

_____ **8.** What happens when an electron moves from an excited state to a ground state?
(1) The electron enters the nucleus.
(2) The electron moves into a higher energy level.
(3) The electron becomes neutral.
(4) The electron moves closer to the nucleus.

_____ **9.** An electron orbital can be thought of as a
(1) shell.
(2) circle.
(3) well-defined region around the nucleus of an atom.
(4) point in space where an electron can be definitely located.

Electron Configurations

All that can be said about the location of the protons and neutrons in an atom is that they are clustered in the nucleus. However, much more can be said about the locations of the electrons in an atom. The location of each electron can be described in some detail. These details are represented as an electron configuration. An **electron configuration** shows how the electrons are arranged in an atom.

An electron configuration is usually written as a series of numbers and letters. First, a number is assigned to represent the shell that the electron occupies. The shell with the lowest energy level is assigned the number 1, followed by the next as number 2, and so on.

Second, a letter is assigned to represent the shape of the electron's orbit within a particular shell. Four letters are mainly used. These four letters are s, p, d, and f.

Third, another number is assigned to show how many electrons occupy the same shell, in terms of both their energy level and orbital shape.

As an example, consider an atom that has two electrons. Remember that electrons must always occupy the lowest energy level available. Therefore, both electrons are in the

Unit II Atomic Structure *continued*

Notes/Study Ideas/Answers

first shell, which is represented by the number 1. Both electrons travel in a shell whose shape is represented by the letter *s*. The electron configuration for this atom is written as follows.

$$1s^2$$

Notice that the 2, which represents the number of electrons that the atom has, is written as a superscript. The first shell contains a maximum of two electrons. If an atom has three electrons, the third electron must be placed in the second shell. An atom with three electrons has the following electron configuration.

$$1s^2 2s^1$$

Self-Check Write the electron configuration for an atom that has four electrons.

Only two electrons can occupy a shell that has a shape represented by the letter *s*. Any additional electrons must be placed in the shell whose shape is represented by the letter *p*. For example, an atom with five electrons has the following electron configuration.

$$1s^2 2s^2 2p^1$$

The above electron configuration is sometimes abbreviated as 2-3. This abbreviated configuration shows that there are a total of five electrons, with two electrons in the first shell and three electrons in the second shell.

The second shell can hold a maximum of eight electrons. Therefore, an atom with ten electrons can be represented with the configuration 2-8.

If more than ten electrons are present, then the third shell begins to be occupied. The third shell can hold a maximum of 18 electrons. For example, the configuration for an atom with 28 electrons can be written as 2-8-18.

Self-Check Write the electron configuration for an atom that has 15 electrons.

Electron configurations show whether all the electrons are in the ground state. Consider the electron configuration 2-8-12. Both the first and the second shells hold their maximum number of electrons, with two electrons in the first shell and eight electrons in the second shell. Electrons do not begin to occupy the third shell until the first two shells have been filled. The configuration 2-8-12 represents an atom whose electrons are in the ground state.

Copyright © by Holt, Rinehart and Winston. All rights reserved.

Atomic Structure

Unit II Atomic Structure continued

Contrast this to an atom that has the electron configuration 2-7-13. Notice that electrons occupy the third shell even though the second shell contains only seven electrons rather than its maximum number of eight electrons. Therefore, this configuration represents an atom with an electron in an excited state.

Self-Check Write the configuration for an atom that has five electrons, one of which is in an excited state.

REVIEW YOUR UNDERSTANDING

_____ 10. What is the maximum number of electrons that can occupy the second shell?
  (1) two
  (2) eight
  (3) eighteen
  (4) an unlimited number of electrons

_____ 11. Identify the electron configuration for an atom with twenty electrons, all of which are in the ground state.
  (1) 2-8-10
  (2) 2-8-12
  (3) 2-7-11
  (4) 2-18

_____ 12. Which configuration represents an atom that has one or more electrons in an excited state?
  (1) 2-8-18
  (2) 2-8-2
  (3) 2-6-2
  (4) 2-6

_____ 13. Which electron transition represents a gain of energy?
  (1) from 3rd to 2nd shell
  (1) from 3rd to 1st shell
  (1) from 2nd to 3rd shell
  (1) from 2nd to 1st shell

_____ 14. What is the total number of protons in an atom that has the following electron configuration?

  2-8-15

  (1) eight
  (2) ten
  (3) fifteen
  (4) twenty-five

Unit II Atomic Structure continued

Notes/Study Ideas/Answers

Electrons and Light

Energy is required to move an electron to a shell that is farther from the nucleus. In turn, energy is released when an electron drops from a higher shell to one that is closer to the nucleus. This energy is usually released as light.

When an electron moves to a shell that is closer to the nucleus, a specific quantity of energy is emitted. This energy can be used to identify the element. The light energy is analyzed by passing it through a device, such as a prism, that separates the various colors present in the light. This spectrum of colors is called a **bright-line spectrum**. Each element produces a unique bright-line spectrum that is made of a different pattern of colors.

Each color consists of light of a different wavelength, which in turn corresponds to a different quantity of energy. For example, green light has a wavelength around 500 nanometers, while red light has a wavelength of around 700 nm. In addition to colors, another way then to identify an element is to analyze the wavelengths of light that are emitted by excited electrons as they drop to shells that are closer to the nucleus.

Self-Check If excited electrons emit an orange-colored light, what would you predict is the wavelength of this light?

A bright-line spectrum always consists of several wavelengths of light that form a unique pattern for each element. These wavelengths are produced because excited electrons fall through various transition levels before they return to their ground state.

For example, an electron may be excited by an input of energy and move from the second shell to the fifth shell. When energy is no longer supplied, the electron will return to its ground state. The electron can drop from the fifth shell directly back to the second shell. When it does, a specific wavelength of light is emitted.

However, the electron can also first drop from the fifth shell to the fourth shell, then to the third shell, and finally to the second shell. Each drop represents an energy transition. With each transition, a specific wavelength of light is emitted.

An electron may also drop from the fifth shell to the third shell, and then to the second shell. Again, specific wavelengths of light are emitted with each energy transition.

Self-Check Assume that an electron is excited from the first shell to the fourth shell. Show all the energy transitions that are possible as this electron returns to its ground state.

A bright-line spectrum can be used to identify the elements present in an unknown sample.

Unit II Atomic Structure continued

REVIEW YOUR UNDERSTANDING

_____ 15. A bright-line spectrum is produced when
   (1) an electron moves from the ground state to an excited state.
   (2) energy is absorbed by an atom.
   (3) all the various colors combine to form white light.
   (4) light energy is emitted.

_____ 16. Which energy transition will produce a bright-line spectrum?
   (1) from 2nd to 3rd shell
   (2) from 3rd to 2nd shell
   (3) from 1st to 5th shell
   (4) from 1st to 2nd shell

_____ 17. A bright-line spectra is produced when electrons in an excited state move from
   (1) lower to higher energy levels, absorbing energy
   (2) lower to higher energy levels, releasing energy
   (3) higher to lower energy levels, absorbing energy
   (4) higher to lower energy levels, releasing energy

SELF-CHECK ANSWERS FOR UNIT II ATOMIC STRUCTURE

- An atom with 46 protons contains 46 electrons, but one cannot be certain how many neutrons it contains.

- An atom with 46 protons contains 46 electrons, but one cannot be certain how many neutrons it contains.

- The farther away from the nucleus that an electron is, the higher its energy.

- Orange-colored light would be between red (700 nm) and green (500 nm), so a good approximation is in the range of 600 nm–630 nm.

- There are four possible transitions from the fourth energy level to the first energy level:
   Fourth direct to first
   Fourth to third, then second, then first
   Fourth to third, then first
   Fourth to second, then first

Questions for Regents Practice

UNIT II

Holt Chemistry: The Physical Setting

Atomic Structure

PART A

_____ 1. Which subatomic particles are *not* found in the nucleus of an atom?
(1) protons
(2) neutrons
(3) electrons
(4) both protons and neutrons

_____ 2. How is the charge on a proton written?
(1) +1
(2) −1
(3) 0
(4) either +1 or −1

_____ 3. The charge of an atom is neutral because it has an equal number of
(1) protons and neutrons.
(2) protons and electrons.
(3) neutrons and electrons.
(4) electrons in each energy level.

_____ 4. Subatomic particles can pass undeflected through an atom because the volume of an atom is composed of
(1) mostly empty space.
(2) the nucleus.
(3) neutrons.
(4) protons.

_____ 5. According to the current model of the atom, electrons travel around the nucleus in
(1) a random fashion that cannot be described.
(2) well-defined, circular paths.
(3) the highest energy levels that are available.
(4) shells, each with a specific energy level.

_____ 6. An electron that is in the lowest energy state possible is said to be in a(n)
(1) excited state.
(2) ground state.
(3) transition state.
(4) energy-emitting state.

_____ 7. What is the maximum number of electrons that can occupy the first shell?
(1) two
(2) four
(3) six
(4) eight

_____ 8. The arrangement of electrons in an atom is represented by a(n)
(1) bright-line spectrum.
(2) electron configuration.
(3) model showing the electrons and protons embedded in the nucleus.
(4) electron transition state diagram.

_____ 9. A bright-line spectrum is produced by
(1) protons that release energy.
(2) neutrons that absorb light.
(3) electrons that emit light.
(4) electrons that absorb energy.

Name _____ Class _____ Date _____

Unit II Atomic Structure *continued*

PART B-1

_____ 10. What is the total number of protons in an atom that has 24 electrons and 25 neutrons?
  (1) 1
  (2) 24
  (3) 25
  (4) 49

_____ 11. Which configuration can represent an atom with all its electrons in the ground state?
  (1) 2-6-1
  (2) 2-8-2
  (3) 2-8-6-2
  (4) 1-2

_____ 12. Identify the configuration for an atom with an electron in an excited state.
  (1) 2-8-1
  (2) 2-2
  (3) 2-1
  (4) 1-2

_____ 13. Identify the electron configuration that will *not* produce a bright-line spectrum.
  (1) 2-8-2
  (2) 2-6-2
  (3) 2-2-2
  (4) 2-8-8-2

_____ 14. How many energy transitions are possible when an electron drops from the fourth shell to the second shell?
  (1) one
  (2) two
  (3) three
  (4) four

PART B-2

15. Why can the number of protons be determined if the number of electrons in an atom is known?

_____

_____

_____

16. When does an electron in an atom have the lowest energy level?

_____

_____

_____

17. Explain what is wrong with each of the following electron configurations.
a. 3-2-1

_____

_____

_____

b. 2-9-8

_____

_____

_____

18. Why does an electron occupy the second shell before the third shell?

_____

_____

_____

Atomic Structure

Name _____ Class _____ Date _____

Unit II Atomic Structure *continued*

PART C

19. Particles that have the same charge, such as protons, repel one another. Therefore, the protons in an atom should repel one another, breaking apart the nucleus. Suggest a possible explanation as to why nuclei are stable and do not break apart spontaneously.

_____

_____

_____

_____

_____

20. A neon sign actually emits a variety of colors. Explain how this is possible in terms of a bright-line spectrum.

_____

_____

_____

_____

_____

_____

_____

_____

Name _____ Class _____ Date _____

UNIT III
Atomic Mass

Holt Chemistry: The Physical Setting

Obviously, an atom is extremely small. In fact, an atom is so small that, until recently, no one had ever seen one, even with the help of the most powerful microscopes. To get an idea of how small an atom is, consider that a copper penny contains about 30,000,000,000,000,000,000,000 (3×10^{22}) copper atoms. Each copper atom weighs only about .000 000 000 000 000 000 000 000 1 gram (1×10^{-25} g). The standard units used to measure mass, even one as small as the picogram (10^{-25} g), are not useful when dealing with atoms. As a result, scientists had to develop a more useful unit when working with the masses of atoms.

Atomic Mass Units

Scientists express atomic mass in a unit that has two names. One is called the Dalton, in honor of John Dalton who first proposed the atomic theory. The other, which is used in this book, is called the **atomic mass unit** (amu). **Atomic mass** can be defined as the mass of an atom expressed in atomic mass units.

At first, scientists did not agree on how to define an atomic mass unit. In 1962, they finally came to an agreement. An atomic mass unit was defined as exactly one-twelfth of the mass of a carbon atom that contains six protons and six neutrons in its nucleus.

As a result of defining the atomic mass unit in this way, each proton has a mass of 1 amu, and each neutron has a mass of 1 amu. Scientists did not take electrons into consideration when they defined the atomic mass unit because electrons have very little mass. The mass of an electron is approximately 0.0005 amu. This means that about 2000 electrons are needed to equal the mass of either one proton or one neutron. Even the largest atoms have only slightly more than 100 electrons. As a result, the mass contributed by electrons is not considered when calculating atomic masses.

What You'll Need to Learn

This topic is part of the Regents Curriculum for the Physical Setting Exam.
Standard 4, Performance Indicator 3.1: *Explain the properties of materials in terms of the arrangement and properties of the atoms that compose them.*

What Terms You'll Need to Know

atomic mass
atomic mass unit
atomic number
average atomic mass
isotope
mass number

Where You Can Learn Even More:

Holt Chemistry: The Physical Setting
Chapter 3: Atoms and Moles
Chapter 7: The Mole and Chemical Composition (Section 2)

Properties of the Subatomic Particles

Particle	Charge	Mass
Proton	+1	1 amu
Neutron	0	1 amu
Electron	−1	negligible

Unit III Atomic Mass continued

Review Your Understanding

_____ 1. Which subatomic particle has a +1 charge and a mass of 1 amu?
(1) proton
(2) electron
(3) neutron
(4) neutron and proton, both

_____ 2. What is the mass of an atom that contains 29 protons and 35 neutrons?
(1) 29 amu
(2) 35 amu
(3) 64 amu
(4) 64 g

_____ 3. Which of the following accounts for very little of an atom's mass?
(1) its electrons
(2) its nucleus
(3) its protons
(4) its neutrons

_____ 4. Identify the element that was chosen as the standard to define the atomic mass unit.
(1) oxygen
(2) hydrogen
(3) gold
(4) carbon

_____ 5. An electron has a charge of
(1) −1 and the same mass as a proton.
(2) +1 and the same mass as a proton.
(3) −1 and a smaller mass than a proton.
(4) +1 and a smaller mass than a proton.

Atomic Number and Mass Number

All atoms consist of protons and electrons. With only one exception, all atoms also consist of neutrons. How then do the atoms of one element differ from those of another element? The answer can be found by looking closely at the nucleus where the protons and neutrons are located.

Elements differ from one another in the number of protons that their atoms contain. The number of protons that an atom has in its nucleus represents its **atomic number**. For example, the atomic number of hydrogen is 1 because all hydrogen atoms

Unit III Atomic Mass continued

have one proton in their nuclei. With two protons in their nuclei, all helium atoms have an atomic number of 2.

No two elements can have the same atomic number because each element has a unique number of protons in the nucleus of its atoms. As an example, any atom whose atomic number is 26 must be an iron atom. Notice that all atomic numbers are whole numbers because an atom cannot have a part of a proton. To date, scientists have identified 113 elements, whose atomic numbers range from 1 to 114.

> **Self-Check** What is the atomic number of an element whose atoms contain 18 protons and 18 neutrons in their nucleus?

Knowing the atomic number of an element also reveals how many electrons are present in an atom of that element. Recall that atoms are neutral because the number of protons they have is equal to the number of electrons they have. Therefore, an element whose atomic number is 31 must have 31 protons in its nucleus and 31 electrons orbiting its nucleus.

To determine how many neutrons are present in an atom, you need to know the element's **mass number**. The mass number is the sum of the numbers of protons and neutrons in the nucleus of an atom. If you know both the atomic number and mass number, you can then calculate how many neutrons, protons, and electrons are present in an atom.

> **Self-Check** What subatomic particles are responsible for the mass number of an atom?

Consider an atom of oxygen whose atomic number is 8 and whose mass number is 16.

atomic number = number of protons = number of electrons

Therefore, this oxygen atom has 8 protons and 8 electrons. Next, calculate how many neutrons are present in an atom of this element.

mass number = number of protons and neutrons
atomic number = number of protons

The mass number minus the atomic number gives the number of neutrons. Therefore, this oxygen atom has 16 minus 8, or 8 neutrons in its nucleus.

Notes/Study Ideas/Answers

Unit III Atomic Mass continued

REVIEW YOUR UNDERSTANDING

_____ 6. Which of the following can be determined by knowing an element's atomic number?
(1) its mass number
(2) the number of neutrons in an atom of this element
(3) the number of protons, only, in an atom of this element
(4) both the number of protons and electrons in an atom of this element

_____ 7. What is the atomic number of an element whose atoms have 21 protons and 23 neutrons in their nucleus?
(1) 2
(2) 21
(3) 23
(4) 44

_____ 8. What is the mass number of an element whose atoms have 82 protons and 125 neutrons in their nucleus?
(1) 43
(2) 82
(3) 125
(4) 207

_____ 9. How many electrons are present in an atom of an element whose atomic number is 47 and whose mass number is 108?
(1) 47
(2) 61
(3) 108
(4) 155

_____ 10. Identify the particles that contribute to an element's mass number.
(1) protons, only
(2) neutrons, only
(3) protons and electrons
(4) protons and neutrons

Isotopes and Atomic Symbols

All atoms of an element have the same atomic number because they have the same number of protons in their nucleus. However, not all atoms of an element have the same mass number. Atoms of an element can differ in the number of neutrons

Unit III Atomic Mass continued

they have in their nucleus. Atoms of an element that differ in their number of neutrons are called **isotopes**.

Carbon, for example, has three isotopes. All three isotopes have an atomic number of 6. Therefore, carbon atoms have six protons in their nucleus and six electrons orbiting the nucleus. One carbon isotope has six neutrons in its nucleus. With six protons and six neutrons, this carbon isotope has a mass number of 12.

Another carbon isotope has seven neutrons in its nucleus, giving it a mass number of 13. The third carbon isotope has eight neutrons, and therefore it has a mass number of 14.

Self-Check Calculate the mass numbers for two isotopes that both have 20 protons. One isotope has 20 neutrons, while the other isotope has 22 neutrons.

Although different elements always have different atomic numbers, they can have the same mass number. For example, potassium has the atomic number 19. With 19 protons, a potassium isotope that has 21 neutrons has a mass number of 40.

19 protons + 21 neutrons = mass number of 40

Now consider calcium whose atomic number is 20. With 20 protons, a calcium isotope that has 20 neutrons also has a mass number of 40.

20 protons + 20 neutrons = mass number of 40

Scientists use symbols to show an element's atomic number and mass number. Each element has a symbol, which is represented by one or two letters. For example, the symbol for calcium is Ca.

The atomic number of an element is always written on the lower left side of the symbol. Therefore, the symbol for calcium is written as follows.

$$_{20}Ca$$

The mass number of an element is always written on the upper left side of the symbol. Therefore, the symbol of a calcium isotope with the mass number 40 is written as follows.

$$^{40}Ca$$

Both the atomic number and mass number may be written with the symbol as shown below.

$$^{40}_{20}Ca$$

By examining the symbol shown above, you should be able to determine that it represents a calcium atom that has 20 protons, 20 electrons, and 20 neutrons.

Notes/Study Ideas/Answers

Copyright © by Holt, Rinehart and Winston. All rights reserved.

Atomic Mass

Unit III Atomic Mass continued

The symbol for a calcium atom that has 21 neutrons is written as follows.

$$^{41}_{20}Ca$$

Self-Check Write the symbol for a silver (Ag) atom that has 47 protons and 61 neutrons.

Sometimes, scientists use another method to represent an element. In this case, the name or the symbol of an element is written, followed simply by its mass number. For example, the calcium isotope whose mass number is 40 is written as calcium-40 or Ca-40. Notice that this method does not tell you how many protons are present in the atom. All you know is that a calcium-40 atom contains a total of 40 protons and neutrons. You could check a reference source to find out calcium's atomic number. Once you know its atomic number, then you could calculate how many protons, neutrons, and electrons are in an atom of calcium-40.

REVIEW YOUR UNDERSTANDING

_____ 11. All isotopes of the same element contain the same number of
   (1) protons, only.
   (2) electrons, only.
   (3) both protons and electrons.
   (4) neutrons, only.

_____ 12. All atoms of oxygen must have the same
   (1) atomic number.
   (2) mass number.
   (3) atomic mass.
   (4) number of neutrons.

_____ 13. How many electrons are present in an atom of $^{133}_{55}Cs$?
   (1) 55
   (2) 78
   (3) 133
   (4) 188

_____ 14. Which represents a pair of isotopes of the same element?
   (1) $^{10}_{5}X$ and $^{10}_{6}X$
   (2) $^{25}_{12}X$ and $^{25}_{12}X$
   (3) $^{65}_{30}X$ and $^{67}_{30}X$
   (4) $^{136}_{56}X$ and $^{138}_{57}X$

Unit III Atomic Mass continued

_____ **15.** Which is the correct symbol for an isotope of silicon (Si) that has 14 protons and 16 neutrons?
   (1) $^{14}_{16}Si$
   (2) $^{14}_{30}Si$
   (3) $^{30}_{14}Si$
   (4) $^{16}_{14}Si$

Average Atomic Mass

Isotopes of the same element have different atomic masses because they have different numbers of neutrons in their nucleus. For example, the atomic mass of copper-63, rounded to a whole number, is 63 amu, while the atomic mass of copper-65, also rounded to a whole number, is 65 amu.

The **average atomic mass** of these two isotopes, however, is not 64 amu. The reason is that these two isotopes are not present in equal amounts in nature. Rather, copper-63 accounts for 69.17% of all copper atoms. Copper-65 makes up the remaining 30.83%.

To determine the average atomic mass of an element, you must take into account two factors: (1) the atomic mass of each isotope, and (2) their relative abundance. The average atomic mass then is a weighted average.

In calculating this weighted average, you must first divide the percentage value by 100 to change it to a fraction. For example, the abundance of copper-63 expressed as a fraction is 69.17/100 = 0.6917. The abundance of copper-65 is 30.83/100 = 0.3083.

> **Self-Check** What two values are needed to calculate an average atomic mass?

The average atomic mass of a copper atom is calculated as follows.

$$(0.6917 \times 63 \text{ amu}) + (0.3083 \times 65 \text{ amu}) = 63.62 \text{ amu}$$

The average atomic mass of a copper atom is closer to copper-63 than it is to copper-65. This result is expected when you consider that copper-63 makes a larger contribution to the average because it represents 69.17% of all copper atoms.

Notes/Study Ideas/Answers

Unit III Atomic Mass continued

Notes/Study Ideas/Answers

REVIEW YOUR UNDERSTANDING

_____ 16. The average atomic mass of a lead atom is 207.2 amu. Which isotope of lead is likely to be the most abundant?
  (1) lead-204
  (2) lead-206
  (3) lead-207
  (4) lead-208

_____ 17. To calculate the average atomic mass of an element, which of the following must you know?
  (1) the atomic number and the atomic mass of each isotope
  (2) the atomic mass and the symbol of each isotope
  (3) the number of atoms present in each isotope
  (4) the atomic mass and relative abundance of each isotope

_____ 18. The average atomic mass of an element is
  (1) always rounded to a whole number.
  (2) a weighted average.
  (3) usually closest to the value of the isotope with the greatest atomic mass.
  (4) calculated by averaging the values of the atomic masses of all the isotopes.

_____ 19. In calculating an average atomic mass, the relative abundance of an isotope is expressed as a(n)
  (1) whole number.
  (2) atomic number.
  (3) fraction.
  (4) atomic mass value.

ANSWERS TO SELF-CHECK QUESTIONS

- The atomic number is 18.

- Protons and neutrons make up an atom's mass number.

- The mass number of the isotope with 20 neutrons is 40. The mass number of the isotope with 22 neutrons is 42.

- $^{108}_{47}Ag$

- The two values required to calculate the average atomic mass of an element are the atomic mass and the relative abundance of each isotope.

Atomic Mass

Name _____ Class _____ Date _____

UNIT III

Questions for Regents Practice

Holt Chemistry: The Physical Setting

Atomic Mass

PART A

_____ 1. Which subatomic particle has a 0 charge and a mass of 1 amu?
   (1) proton
   (2) electron
   (3) neutron
   (4) both a neutron and a proton

_____ 2. The mass number of an atom is always equal to the total number of its
   (1) protons, only.
   (2) electrons, only.
   (3) neutrons, only.
   (4) protons plus neutrons.

_____ 3. The number of neutrons in an atom can be calculated by
   (1) adding its atomic number and mass number.
   (2) examining its electron configuration.
   (3) subtracting the mass number from the atomic number.
   (4) subtracting the atomic number from the mass number.

_____ 4. An atom contains 21 protons and 24 neutrons. What is this atom's mass number?
   (1) 3
   (2) 21
   (3) 24
   (4) 45

_____ 5. All isotopes of a given element must have the same
   (1) mass number.
   (2) atomic number.
   (3) number of neutrons.
   (4) atomic mass.

_____ 6. An atom of potassium-39
   (1) has an atomic number of 39.
   (2) contains 39 protons.
   (3) contains 39 protons and neutrons.
   (4) cannot have any other isotopes.

_____ 7. Which number is written to the upper left of an element's symbol?
   (1) atomic number
   (2) mass number
   (3) atomic mass
   (4) number of neutrons it contains

_____ 8. The average atomic mass of an element is defined as the weighted average of that element's
   (1) most abundant isotopes.
   (2) least abundant isotopes.
   (3) total number of isotopes.
   (4) atomic numbers.

_____ 9. The average atomic mass of an element is closest to the isotope of that element that
   (1) is most abundant.
   (2) is least abundant.
   (3) has the greatest atomic mass.
   (4) has the smallest atomic mass.

Unit III Atomic Mass continued

PART B-1

_____ 10. What is the total number of neutrons in an element with an atomic number of 28 and a mass number of 58?
  (1) 28
  (2) 30
  (3) 58
  (4) 86

_____ 11. The nucleus of an atom of $^{35}_{17}Cl$ contains
  (1) 17 protons and 18 neutrons.
  (2) 17 protons and 35 neutrons.
  (3) 35 protons and 17 neutrons.
  (4) 35 protons and 18 neutrons.

_____ 12. Which represents a pair of isotopes?
  (1) oxygen-16 and $^{16}_{8}O$
  (2) aluminum-27 and $^{13}_{27}Al$
  (3) carbon-14 and nitrogen-14
  (4) calcium-40 and $^{41}_{20}Ca$

_____ 13. All atoms of $^{20}_{10}Ne$ have
  (1) 10 protons.
  (2) 20 neutrons.
  (3) 20 protons.
  (4) 30 protons and neutrons.

_____ 14. Element X has three isotopes. They exist in nature as 75% of the isotope ^{20}X, 18% of the isotope ^{18}X, and 7% of the isotope ^{17}X. Its average atomic mass is closest to a value of
  (1) 17 amu.
  (2) 18 amu.
  (3) 20 amu.
  (4) 75 amu.

PART B-2

15. Compare and contrast the charges and masses of a proton, neutron, and electron.

_____

_____

_____

16. Neutrons contribute to the mass of an atom. However, explain why you cannot determine how many neutrons are present in an atom even if you know its mass number.

_____

_____

_____

17. Prepare a table to show how many protons, electrons, and neutrons are present in two isotopes of barium (Ba), whose atomic number is 56. One isotope is Ba-130, and the other isotope is Ba-137.

_____

_____

18. Calculate the average atomic mass for the element gallium. This element has two isotopes, with 60% having a mass of 69 amu and 40% having a mass of 71 amu.

_____

_____

_____

Unit III Atomic Mass *continued*

PART C

19. For hundreds of years, people searched for ways to turn various metals, such as lead, into gold. How would you change the structure of an atom of $^{208}_{82}$Pb (lead) into $^{197}_{79}$Au (gold)?

20. The oxygen atoms in the air exist as three isotopes. Oxygen's composition in nature consists of 99.76% of atoms with a mass of 16 amu, 0.038% with a mass of 17 amu, and 0.20% with a mass of 18 amu. Calculate the average atomic mass of an oxygen atom.

Name _____ Class _____ Date _____

UNIT IV
The Periodic Table

Holt Chemistry: The Physical Setting

By the 1860s, scientists had identified about 60 elements. They tried to arrange them in some logical fashion. This would make them easier to study and perhaps point out similarities and differences among the elements. One of the first attempts to organize the elements was to arrange them according to increasing atomic mass. When this was done, a pattern emerged. The physical and chemical properties of the elements seemed to repeat every eight elements. In 1869, a Russian chemist named Dmitri Mendeleev examined the known elements more closely. His observations led to the development of the modern periodic table.

The Modern Table

Mendeleev arranged the elements in a row based on increasing atomic mass. He started a new row each time he noticed that the chemical properties of the elements repeated. Mendeleev placed elements in the new row directly below elements with similar chemical properties in the preceding row. As a result, Mendeleev developed a table of the elements, consisting of rows and columns.

However, Mendeleev's table of elements presented a problem. There were several places where an element in a column did not share the same properties as the other elements in that column. Mendeleev was not able to solve this problem.

In the early 1900s, an English chemist named Henry Moseley figured out the problem with Mendeleev's table. Recall that Mendeleev, like others before him, had set up his table based on increasing atomic mass. However, Moseley set up his table of elements based on increasing atomic number. When the elements were arranged this way, the problem posed by Mendeleev's table disappeared.

Today, a periodic table lists 113 elements, all arranged according to increasing atomic number. A periodic table is not just an organized list of all the known elements. Rather, a periodic table contains a good deal of information about each element. Different tables contain different kinds of information. However, most periodic tables contain some basic information that can be used to describe the element in more detail, predict how an element will react, and solve mathematical problems involving the element. To do all this, you must understand what information is contained on a periodic table.

What You'll Need to Learn
This topic is part of the Regents Curriculum for the Physical Setting Exam.
Standard 4, Performance Indicator 3.1: *Explain the properties of materials in terms of the arrangement and properties of the atoms that compose them.*

What Terms You'll Need to Know
alkali metal
alkaline-earth metal
ductile
electronegativity
group
halogen
ionization energy
malleable
metal
metalloid
noble gas
period
valence electron

Where You Can Learn Even More:
Holt Chemistry: The Physical Setting
Chapter 4: The Periodic Table

Unit IV The Periodic Table continued

As an example, consider the element carbon as it appears on many periodic tables.

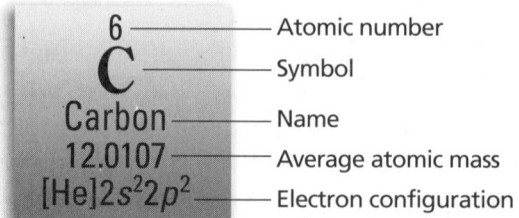

Notice that a periodic table lists each element according to its symbol, in this case C for carbon. In addition, a periodic table also lists an element's atomic number. You learned that an atomic number represents the number of protons an element has in each of its atoms. Therefore, you can tell from a periodic table that a carbon atom has six protons. You also learned that an atom is neutral. Therefore, a carbon atom also has six electrons.

Next, notice that a periodic table lists an element's average atomic mass. With some exceptions, these values are not whole numbers. Recall that an average atomic mass is based on the mass and relative abundance of all the isotopes of that element.

You can use the average atomic mass to tell more about an atom of one of the element's isotopes. However, you must first round the value to the nearest whole number. For carbon, the value is rounded from 12.0107 to 12. Now, you can determine how many neutrons are in an atom of carbon-12. Take the mass number and subtract the atomic number, and the result is the number of neutrons in a carbon-12 atom. In this case, $12 - 6 = 6$ neutrons.

Self-Check Use a periodic table to determine how many protons, electrons, and neutrons are present in an atom of sulfur-32, whose symbol is S.

Another piece of information contained on most periodic tables is the element's electron configuration. The electron configuration for carbon can be listed in several ways. It may be written as $1s^2 2s^2 2p^2$. Carbon's electron configuration can also be listed as $[He]2s^2 2p^2$. This method is a short-hand way of saying that carbon's electron configuration is the same as that of helium, whose symbol is He, plus $2s^2 2p^2$.

If you examine a periodic table that is supplied with a Regents exam, you will notice that carbon's electron configuration is presented in still another way: 2-4. This way of presenting the electron configuration tells you that a carbon atom contains two electrons in its first shell and four electrons in its second shell.

Name _____ Class _____ Date _____

Unit IV The Periodic Table continued

REVIEW YOUR UNDERSTANDING

Notes/Study Ideas/Answers

_____ 1. What basis did Mendeleev use for organizing his periodic table?
(1) atomic numbers
(2) atomic masses
(3) the symbols for the elements
(4) electron configurations

_____ 2. The elements in the present periodic table are arranged according to their
(1) atomic numbers.
(2) mass numbers.
(3) average atomic masses.
(4) electron configurations.

_____ 3. What is the electron configuration for an atom of chlorine (Cl)?
(1) 2-8
(2) 2-8-2
(3) 2-8-6
(4) 2-8-7

_____ 4. How many neutrons are present in an atom of magnesium-26? The symbol for magnesium is Mg.
(1) 2
(2) 12
(3) 14
(4) 26

_____ 5. Which element has the greatest average atomic mass?
(1) potassium (K)
(2) mercury (Hg)
(3) iron (Fe)
(4) sodium (Na)

Periods and Groups

A horizontal row on a periodic table is called a **period**. The present periodic table contains seven periods, numbered 1 through 7. Elements in the same period have the same number of occupied shells. For example, all elements in Period 3 have atoms whose electrons occupy three shells. If you check a periodic table, you will see that sodium (Na), the first element in Period 3, has the electron configuration 2-8-1. Argon (Ar), the last element in Period 3, has the configuration 2-8-8.

Copyright © by Holt, Rinehart and Winston. All rights reserved.

The Periodic Table — Focus On Regents Exam

Unit IV The Periodic Table *continued*

The correlation between period number and occupied shells applies to all seven periods. Therefore, an atom whose electron configuration is 2-8-8-1 belongs to Period 4 because its electrons occupy four shells.

> **Self-Check** To what period does an element whose atoms have the configuration 2-8-18-8-1 belong?

A vertical column on a periodic table is called a **group**. The present periodic table contains eighteen groups, numbered 1 through 18. Some periodic tables also label groups with Roman numerals and letters, including A, B, and O. This is the older method once used for identifying a group. For example, Group 1 was referred to as IA, while Group 18 was known as Group O.

Knowing an element's group number reveals more information about that element's activity than knowing its period number. The reason is that elements in the same group have the same number of electrons in their outermost shell. Electrons in the outermost shell are known as **valence electrons**. Elements with the same number of valence electrons tend to react in the same way.

The elements in Group 1 have one valence electron in their atoms. These elements react in a similar manner. In contrast, the atoms of elements in Group 18 have eight valence electrons. Elements in Group 18 react in a similar manner, but very differently from those in Group 1.

> **Self-Check** Check a periodic table to determine how many valence electrons the atoms of the elements in Group 14 and Group 17 have.

Several groups have names. For example, elements in Group 1 are known as **alkali metals**. These elements are so named because they are metals that react readily with water to produce an alkaline solution. The alkali metals are usually stored in oil to keep them from reacting with the water or oxygen in air. All alkali metals are so soft that they can be cut with knife.

Group 2 elements are called **alkaline-earth metals**. These metals are just slightly less reactive than the alkali metals of Group 1.

Group 17 elements are known as **halogens**. Like Group 1 and Group 2 elements, the halogens are very reactive. However, the halogens are not metals.

Group 18 elements are called **noble gases**. Members of this group are unreactive. At one time, noble gases were called inert gases because they were thought to be completely unreactive. However, scientists have been able to make some of the noble gases react with other elements.

Name _____ Class _____ Date _____

Unit IV The Periodic Table continued

REVIEW YOUR UNDERSTANDING

_____ 6. Which element is a halogen?
   (1) S
   (2) Na
   (3) I
   (4) H

_____ 7. Which pair of elements belongs to the same period?
   (1) K and Cs
   (2) H and He
   (3) C and Ca
   (4) Fe and Fr

_____ 8. All elements in Group 2
   (1) are called alkali metals.
   (2) have atoms whose electrons occupy two shells.
   (3) have two valence electrons.
   (4) are unreactive.

_____ 9. How many valence electrons are present in an element whose atom has the electron configuration 2-8-8-1?
   (1) 1
   (2) 2
   (3) 8
   (4) 19

Classifying Elements

A group contains elements whose atoms have the same number of valence electrons. A period contains elements whose atoms have the same number of occupied shells. However, elements from different groups and periods can be classified into broader categories. Each category contains elements that share certain properties.

One category includes all the elements that are metals. Some periodic tables indicate the elements that are metals, usually by shading them the same color. If you check such a periodic table, you will notice that most elements are metals. You will also notice that the metals are located on the left side and in the center of a periodic table. A few are found toward the right side.

Metals share several properties. The one property that all metals have is their ability to conduct electricity. Metals are also good conductors of heat. Metals exhibit other properties, some of which may be displayed by elements that are not metals. For example, metals are **ductile**, which means that they can be squeezed to form a wire. Metals are also **malleable**, which means that they can be hammered or rolled into sheets.

Notes/Study Ideas/Answers

Copyright © by Holt, Rinehart and Winston. All rights reserved.

Unit IV The Periodic Table *continued*

Self-Check What property distinguishes metals from the other elements?

If you check a periodic table, you will notice that some of the elements in Period 6 and Period 7 are placed at the bottom of a periodic table. This is done simply to keep the periodic table conveniently narrow. All the elements at the bottom are metals.

However, not all the elements in Period 6 are metals. Two of them are nonmetals, which make up another category of elements. The two nonmetals in Period 6 are astatine (At), a halogen, and radon (Rn), a noble gas.

Nonmetals are located toward the right side of a periodic table. Nonmetals include all the halogens (Group 17) and all the noble gases (Group 18). In addition, some elements in Groups 14, 15, and 16 are nonmetals.

A third category of elements is located between the metals on the left side and the nonmetals on the right side of a periodic table. These elements are known as **metalloids** because they have some properties of metals and other properties of nonmetals.

Some periodic tables use a zigzag border to separate the metals from the nonmetals. Elements located along this border are the metalloids. Perhaps the best-known metalloid is silicon (Si), which is used to make computer chips. In addition to silicon, five other elements are categorized as metalloids.

Self-Check Check a periodic table and identify the six metalloids.

Another way to classify the elements is as solids, liquids, and gases. At room temperature, most metals are solids, and most nonmetals are either solids or gases. The two exceptions are the metal mercury (Hg) and the nonmetal bromine (Br). Both Hg and Br are liquids at room temperature.

Still another way to classify the elements is to place them in one of two categories: natural elements and synthetic elements. There are 93 natural elements, or ones that are found in nature. Of these, only 90 elements are found on Earth, while the other three natural elements have been detected in stars.

Elements whose atoms have more than 92 protons in their nucleus are called synthetic elements. These elements have been made with the use of special equipment known as particle accelerators.

One natural element that is a nonmetal is usually placed in a category by itself. This element is hydrogen. Hydrogen exists as three isotopes: hydrogen-1, hydrogen-2, and hydrogen-3.

Name _____ Class _____ Date _____

Unit IV The Periodic Table continued

Hydrogen-1 is a unique element because it consists of just one proton and one electron. Of all the elements, hydrogen-1 is the only isotope that does not have any neutrons.

Notes/Study Ideas/Answers

REVIEW YOUR UNDERSTANDING

_____ 10. Identify the element that is a metalloid.
   (1) H
   (2) C
   (3) B
   (4) Sn

_____ 11. Which element is considered malleable?
   (1) chlorine (Cl)
   (2) xenon (Xe)
   (3) oxygen (O)
   (4) magnesium (Mg)

_____ 12. Most elements on a periodic table are
   (1) metals.
   (2) nonmetals.
   (3) metalloids.
   (4) gases.

_____ 13. The element whose atoms have the electron configuration 2-8-8 is a(n)
   (1) synthetic element.
   (2) nonmetal.
   (3) metal.
   (4) metalloid.

_____ 14. Which group on a periodic table contains a metalloid?
   (1) Group 1
   (2) Group 2
   (3) Group 8
   (4) Group 16

Periodic Trends

A periodic table reveals trends in the physical and chemical properties of elements. These trends can be found by examining either a period or a group of elements. These trends can be explained in terms of the atomic structures, in particular the electron configurations, of the elements.

Atomic radius is one trend that can be identified on a periodic table. Atomic radius refers to the size of an atom. An atom's size is very difficult to measure with any precision

Copyright © by Holt, Rinehart and Winston. All rights reserved.
The Periodic Table

because its electrons do not move in well-defined paths. Nonetheless, scientists can assign values to the radii of atoms.

Atomic radius increases as you move down a group. This increase is mainly the result of the addition of another electron shell each time you move from one element down to the next in a group.

Atomic radius decreases as you move left to right across a period. This decrease is mainly the result of the addition of another proton and electron to an atom each time you move from one element across to the next in a period. The addition of these subatomic particles increases the attractive force between the nucleus and the electrons. As a result, the orbiting electrons are pulled closer to the nucleus.

> **Self-Check** How does atomic radius change as you move down a group and across a period?

Electronegativity is another trend that can be observed in a periodic table. Electronegativity is a value assigned to an element to indicate how strongly its atoms attract electrons.

Electronegativity decreases as you move down a group. This decrease is mainly the result of the greater distance between the valence electrons and the nucleus of an atom each time you move from one element down to the next in a group. Because of this greater distance, the nucleus of an atom has less ability to attract electrons.

Electronegativity increases as you move left to right across a period. This increase is mainly the result of the stronger attraction the nucleus has for its electrons each time you move from one element across to the next in a period.

> **Self-Check** How does electronegativity change as you move down a group and across a period?

Ionization energy is another periodic trend. Ionization energy is the energy required to remove an electron from an atom. The energy required to remove the first electron from an atom is known as the *first ionization energy*.

Ionization energy decreases as you move down a group. This decrease is mainly the result of the greater distance between the valence electrons and the nucleus of an atom each time you move from one element down to the next. Because of this greater distance, less energy is required to remove an electron, which is usually a valence electron in the outermost shell.

Ionization energy increases as you move across a period. This increase is a result of the stronger attraction the nucleus

Unit IV The Periodic Table continued

has for its electrons each time you move from one element across to the next in a period. Because of this stronger attraction, more energy is required to remove an electron from an atom.

Self-Check How does ionization energy change as you move down a group and across a period?

Still other trends can be observed on a periodic table, including metallic properties. Generally, metallic properties increase as you move down a group and decrease as you move left to right across a period.

The following illustrations summarize the periodic trends as seen in atomic radius, electronegativity, and ionization energy.

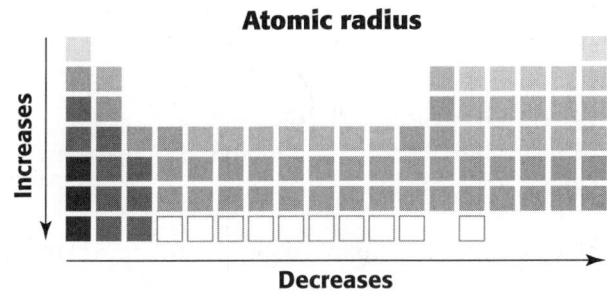

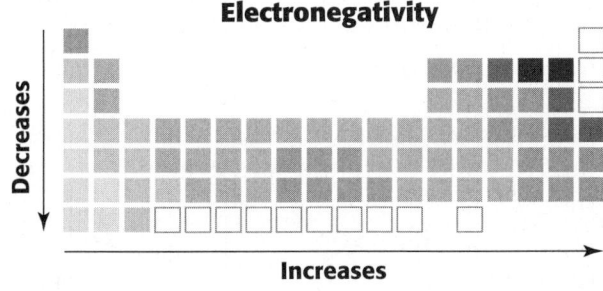

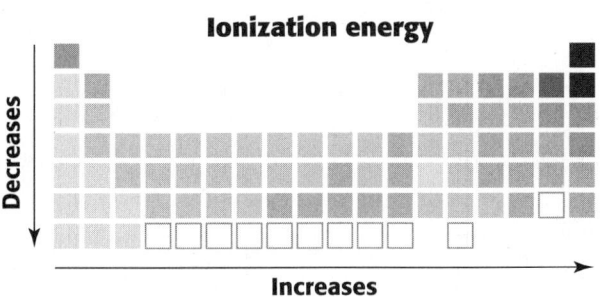

Name _____ Class _____ Date _____

Unit IV The Periodic Table continued

Notes/Study Ideas/Answers

REVIEW YOUR UNDERSTANDING

_____ **15.** Which element in Period 3 has an atom with the smallest atomic radius?
  (1) aluminum (Al)
  (2) sodium (Na)
  (3) chlorine (Cl)
  (4) phosphorus (P)

_____ **16.** Which element in Group 1 has the atom that requires the *least* amount of energy to remove one of its valence electrons?
  (1) lithium (Li)
  (2) sodium (Na)
  (3) potassium (K)
  (4) cesium (Cs)

_____ **17.** As you proceed down a group,
  (1) atomic radius increases.
  (2) electronegativity increases.
  (3) ionization energy increases.
  (4) atomic radius decreases.

_____ **18.** Potassium (K) has a larger atomic radius than sodium (Na) because potassium
  (1) is not in the same group.
  (2) has more electron shells that are occupied.
  (3) has fewer electrons that sodium.
  (4) is a noble gas.

_____ **19.** In Period 2, as the elements are considered in moving from left to right, there is a general increase in their
  (1) metallic nature.
  (2) atomic size.
  (3) electronegativity values.
  (4) ability to conduct heat.

Unit IV The Periodic Table continued

ANSWERS TO SELF-CHECK QUESTIONS

- A sulfur-32 atom consists of 16 protons, 16 electrons, and 16 neutrons.

- This element belongs to Period 5.

- Group 14 elements have atoms with four valence electrons. Group 17 elements have atoms with seven valence electrons.

- All metals are excellent conductors of electricity.

- The six metalloids are boron (B), silicon (Si), germanium (Ge), arsenic (As), antimony (Sb), and tellurium (Te).

- Atomic radius increases as you move down a group and decreases as you move left to right across a period.

- Electronegativity decreases as you move down a group and increases as you move left to right across a period.

- Ionization energy decreases as you move down a group and increases as you move left to right across a period.

Notes/Study Ideas/Answers

Holt Chemistry: The Physical Setting

UNIT IV
Questions for Regents Practice

The Periodic Table
PART A

_____ 1. The elements in a periodic table are arranged in order of increasing
(1) atomic number.
(2) mass number.
(3) neutron number.
(4) metallic properties.

_____ 2. Which element has physical and chemical properties most similar to those of oxygen (O)?
(1) nitrogen (N)
(2) neon (Ne)
(3) sulfur (S)
(4) osmium (Os)

_____ 3. A horizontal row on a periodic table is called a
(1) category.
(2) period.
(3) group.
(4) periodic trend.

_____ 4. Which period includes elements whose atoms have three shells occupied by electrons?
(1) Period 1
(2) Period 2
(3) Period 3
(4) Period 7

_____ 5. Which element is a halogen?
(1) potassium (K)
(2) oxygen (O)
(3) hydrogen (H)
(4) iodine (I)

_____ 6. Which elements are liquids at room temperature?
(1) hydrogen (H) and helium (He)
(2) bromine (Br) and mercury (Hg)
(3) chlorine (Cl) and iodine (I)
(4) sodium (Na) and potassium (K)

_____ 7. The two metalloids in Period 4 are
(1) titanium (Ti) and zirconium (Zr).
(2) antimony (Sb) and tellurium (Te).
(3) germanium (Ge) and arsenic (As).
(4) boron (B) and silicon (Si).

_____ 8. Which represents a pair of elements that are malleable?
(1) hydrogen (H) and helium (He)
(2) fluorine (F) and chlorine (Cl)
(3) iron (Fe) and gold (Au)
(4) any two elements in Group 16

_____ 9. The ability of an atom to attract electrons is called its
(1) ionization energy.
(2) electronegativity.
(3) electron configuration.
(4) atomic radius.

Copyright © by Holt, Rinehart and Winston. All rights reserved.
The Periodic Table
Focus On Regents Exam

Unit IV The Periodic Table continued

PART B-1

_____ 10. The element in Period 3 and Group 1 of a periodic table is classified as a
(1) noble gas.
(2) metalloid.
(3) metal.
(4) synthetic element.

_____ 11. In the ground state, the atoms of the elements in Group 14 all have the same number of
(1) valence electrons.
(2) occupied energy levels.
(3) protons in their nucleus.
(4) electron shells.

_____ 12. As the elements in Period 3 of a periodic table are considered in succession from left to right, there is a decrease in atomic radius with increasing atomic number. This decrease in atomic radius may best be explained by the fact that the
(1) number of protons increases, and the number of shells of electrons remains the same.
(2) number of protons increases, and the number of shells of electrons increases.
(3) number of protons decreases, and the number of shells of electrons remains the same.
(4) number of protons decreases, and the number of shells of electrons increases.

_____ 13. Compared to the nonmetals in Period 2, the metals in Period 2 generally have larger
(1) ionization energies.
(2) atomic radii.
(3) atomic numbers.
(4) electronegativities.

_____ 14. Which Group 2 element has the lowest first ionization energy?
(1) magnesium (Mg)
(2) beryllium (Be)
(3) barium (Ba)
(4) calcium (Ca)

PART B-2

15. Why is iodine (I) placed just after tellurium (Te) on a periodic table even though the atomic mass of iodine is less than that of tellurium?

_____

_____

_____

16. Boron (B) and carbon (C) are in the same period. Carbon and silicon (Si) are in the same group. Would you expect carbon to have properties similar to those of boron or to those of silicon? Explain your answer.

_____

_____

_____

17. Explain why barium (Ba) is placed in Group 2 and in Period 6 of a periodic table.

_____

_____

_____

Unit IV The Periodic Table *continued*

18. Construct a line graph that illustrates how ionization energy changes as you proceed down Group 18. Refer to *Reference Table S* which lists the first ionization energy values for the Group 18 elements.

_____

_____

PART C

19. The element neon (Ne), which is used in signs, differs from both fluorine (F) and sodium (Na) by one proton each. Sodium, an alkali metal, and fluorine, a halogen, are both extremely reactive. In contrast, neon, which is a noble gas, is highly unreactive. Use the information contained on a periodic table to explain the difference in their chemical properties.

_____

_____

_____

_____

_____

20. Element 118 has yet to be synthesized. In which period and in which group would this element be placed if it is synthesized? Predict how element 118 should behave chemically.

_____

_____

_____

_____

_____

_____

UNIT V
Ionic Compounds

Holt Chemistry: The Physical Setting

All atoms are neutral because their number of protons equals their number of electrons. However, not all atoms are reactive. In fact, some do not react, such as the noble gases, unless certain conditions are met. Those that do react can do so very quickly or do so very slowly. What makes an atom react quickly, slowly, or not at all? The answer can be found by looking at an atom's electron configuration.

Stability and Ions

A clue as to why certain atoms are reactive can be found by looking at atoms that are unreactive—the noble gases. As members of Group 18, atoms of the noble gases have the same number of valence electrons. There is one exception. Helium has only two valence electrons. However, helium also has only one electron shell, or principal energy level. This shell can hold only two electrons. So, helium's only shell is fully occupied with two electrons.

The remaining noble gases in Group 18—from neon to radon—all have eight valence electrons. As a member of Period 2, neon has eight valence electrons in its second shell. As member of Period 6, radon has eight valence electrons in its sixth shell. Eight valence electrons make an atom stable.

Keep in mind that helium's valence shell is fully occupied. If an atom with only one shell, such as helium, contains two electrons, then that atom is also stable.

Atoms other than the noble gases are not stable because they do not have completely filled valence shells. Consider sodium and chlorine, which are both members of Period 3. Sodium has one valence electron, while chlorine has seven valence electrons. These atoms will react so that they have eight valence electrons. However, sodium and chlorine will not react the same way.

Examining a periodic table reveals that sodium's electron configuration is 2-8-1. Sodium can gain seven electrons so that its third shell will have eight electrons. However, sodium can also lose its one valence electron. If this happens, then sodium's second shell becomes its outermost shell. Sodium will then have eight valence electrons and be stable. Losing one electron requires much less energy than gaining seven electrons. Therefore, sodium tends to give up its one valence electron. When sodium gives up its one valence electron, it will have the same configuration as the noble gas neon: 2-8.

What You'll Need to Learn
This topic is part of the Regents Curriculum for the Physical Setting Exam.
Standard 4, Performance Indicator 5.2: *Explain chemical bonding in terms of the behavior of electrons.*

What Terms You'll Need to Know
anion
binary compound
cation
ion
ionic bond
polyatomic ion

Where You Can Learn Even More:
Holt Chemistry: The Physical Setting
Chapter 5: Ions and Ionic Compounds

Ionic Compounds

Unit V Ionic Compounds continued

When sodium loses its valence electron, it also loses its neutrality. The nucleus of a sodium atom still contains 11 protons. However, by giving up its valence electron, sodium now has only 10 electrons. With 11 protons and 10 electrons, sodium now has 11 positive charges and 10 negative charges. Therefore, sodium now has a 1+ charge.

When sodium loses its valence electron and develops a 1+ charge, it is no longer called an atom but an **ion**. An ion is an atom that has either lost or gained one or more electrons and develops a positive or negative charge. An ion with a positive charge is called a **cation**.

The formation of a sodium cation is shown in the following equation.

$$Na + energy \rightarrow Na^+ + e^-$$

Notice that the charge on a sodium cation is written as a superscript. However, the charge is indicated simply as a + because the number 1 is not written in equations. The same applies to an electron with its 1– charge. The energy in the equation above represents the first ionization energy.

> **Self-Check** Write the equation to show how potassium (K) forms a cation.

Metals, such as sodium, tend to form ions by losing electrons. Therefore, nearly all metals form cations. Consider a magnesium (Mg) atom that has two valence electrons as shown by its electron configuration: 2-8-2.

To achieve stability, a magnesium atom must lose its two valence electrons. Its electron configuration will then be the same as that of a neon atom: 2-8. When a magnesium atom loses its two valence electrons, it will also have two fewer electrons than protons. As a result, a magnesium cation has a 2+ charge as shown in the following equation.

$$Mg + energy \rightarrow Mg^2 + 2e^-$$

Notice that the charge, 2+, is written as a superscript to indicate the charge on a magnesium cation. Notice also that a number 2 is written as a coefficient to indicate that two electrons have been given up. Again, the energy in the equation above represents the ionization energy that is required to remove the two valence electrons from a magnesium atom.

> **Self-Check** Write the equation to show how aluminum (Al) forms a stable cation. Hint: Check a periodic table for aluminum's electron configuration

Name _____ Class _____ Date _____

Unit V Ionic Compounds continued

Nonmetals tend to form ions differently than metals. The reason is that nonmetals tend to have six or seven valence electrons. Consider the electron configuration of a chlorine atom—2-8-7. To achieve stability, a chlorine atom can either gain one electron or lose its seven valence electrons. Losing seven electrons takes a good deal of energy. Losing one electron, however, actually releases energy, as shown in the following equation.

$$Cl + e^- \rightarrow Cl^- + energy$$

When chlorine gains an electron, the ion that is formed has a 1– charge, as shown in the above equation. An ion with a negative charge is called an **anion**. A chlorine anion has the same electron configuration as an argon atom: 2-8-8.

> **Self-Check** Write the equation to show how oxygen (O) forms an anion. Hint: Check a periodic table for oxygen's electron configuration.

Besides acquiring a negative charge, something else happens when an atom gains an electron. Consider chlorine as an example. A chlorine atom has 17 protons and 17 electrons. A chorine anion has 17 protons and 18 electrons. With the same number of protons, the nucleus of a chorine anion cannot attract its 18 electrons as strongly as the nucleus of a chlorine atom with its 17 electrons. Therefore, the electrons in the chlorine anion can move farther from the nucleus than they can in a chlorine atom.

As a result, nonmetal elements tend to form anions that are larger than the neutral atoms. In other words, when an atoms gains one or more electrons, it becomes a negative ion and its radius increases.

In contrast, metals tend to form cations that are smaller than the neutral atoms. In other words, when an atom loses one or more electrons, it becomes a positive ion and its radius decreases.

REVIEW YOUR UNDERSTANDING

_____ 1. What occurs when an atom of chlorine forms an ion?
  (1) The chlorine atom gains an electron, and its radius become smaller.
  (2) The chlorine atom gains an electron, and its radius become larger.
  (3) The chlorine atom loses an electron, and its radius become smaller.
  (4) The chlorine atom loses an electron, and its radius become larger.

Notes/Study Ideas/Answers

Name _____ Class _____ Date _____

Unit V Ionic Compounds continued

Notes/Study Ideas/Answers

_____ 2. A strontium (Sr) atom differs from a strontium ion in that the atom has a greater
(1) number of protons.
(2) atomic number.
(3) number of electrons.
(4) number of neutrons.

_____ 3. When calcium (Ca) forms a cation, its electron configuration becomes
(1) identical to that of argon (Ar).
(2) identical to that of all other elements in Period 4.
(3) 2-8-8-2.
(4) identical to any of the noble gases.

_____ 4. The atoms of metals tend to form
(1) anions that have radii larger than the atoms.
(2) anions that have radii smaller than the atoms.
(3) cations that have radii larger than the atoms.
(4) cations that have radii smaller than the atoms.

_____ 5. Which represents the formation of a stable ion by a sulfur (S) atom?
(1) $S + energy \rightarrow S^{2+} + 2e^-$
(2) $S + 2e^- \rightarrow S^{2-} + energy$
(3) $S + e^- + energy \rightarrow S^-$
(4) $S + 2e^- \rightarrow S^{2+} + energy$

Ionic Bonds

When an atom loses one or more electrons, it becomes a cation. When an atom gains one or more electrons, it becomes an anion. Because they have opposite charges, cations and anions should attract one another. This is exactly what happens. A cation and an anion attract each other and join to form an **ionic bond**. An ionic bond is a bond that forms between ions that have opposite charges. One or more electrons must be transferred from one atom to another atom to form an ionic bond.

For example, a sodium atom gives up an electron, which can be picked up by a chlorine atom. The stable Na+ cation and the stable Cl– anion then attract each other. The force of attraction between the 1+ charge on the sodium cation and the 1– charge on the chloride anion creates an ionic bond.

Self-Check Explain how an ionic bond forms between two atoms.

The ionic bond that forms between Na^+ and Cl^- creates a compound called sodium chloride, which is commonly called

Unit V Ionic Compounds continued

table salt. Sodium chloride is only one of thousands of different salts that can be formed between cations and anions that are joined by ionic bonds. The tiniest salt crystal actually consists of many billions of cations and anions held together by ionic bonds.

The name of an ionic compound, such as sodium chloride, is based on the names of the cation and anion that make up the compound. The cation retains the name of its parent atom. For example, Na^+ is called a sodium cation, and Mg^{2+} is called a magnesium cation.

However, the name of the anion is slightly different from its parent atom. For example, Cl^- is called a chloride anion, and S^{2-} is called a sulfide anion. In naming anions, you simply change the name of the parent atom so that it ends in -*ide*. Some changes are more involved than others. For example, oxygen atoms form oxide anions, and nitrogen atoms form nitride anions.

Self-Check Name the compound that is produced when magnesium and oxygen form an ionic bond.

Only one electron is transferred between each sodium atom and each chlorine atom when sodium chloride is formed. However, the formation of an ionic bond does not always involve the transfer of just a single electron.

Consider what happens when magnesium reacts to form an ionic bond with sulfur. Magnesium has two valence electrons, while sulfur has six valence electrons. In this case, an ionic bond is formed when a magnesium atom loses its two valence electrons and a sulfur atom gains those two electrons. The following equations show this process.

formation of the cation: $Mg + energy \rightarrow Mg^{2+} + 2e^-$
formation of the anion: $S + 2e^- \rightarrow S^{2-} + energy$
formation of the ionic bond: $Mg^{2+} + S^{2-} \rightarrow MgS + energy$

There are two important observations to be made. First, energy is released whenever an ionic bond is formed. Second, you can refer to a salt, or any other compound, by either its name or its chemical formula. In this case, the name of the salt is magnesium sulfide, whose formula is written as MgS.

Name _____ Class _____ Date _____

Unit V Ionic Compounds continued

Notes/Study Ideas/Answers

REVIEW YOUR UNDERSTANDING

_____ 6. What must happen to a barium (Ba) atom to form a stable ion?
  (1) The Ba atom loses its two valence electrons to form a cation
  (2) The Ba atom loses its two valence electrons to form an anion.
  (3) The Ba atom adds two electrons to its valence shell.
  (4) The Ba atom gains six electrons.

_____ 7. When an ionic bond forms,
  (1) energy is absorbed.
  (2) energy is released.
  (3) energy can be either absorbed or released.
  (4) atoms of metals tend to bond with atoms of other metals.

_____ 8. Which represents the ion formed when zinc (Zn) loses two electrons?
  (1) Zn
  (2) Zn^+
  (3) Zn^{2+}
  (4) Zn^{2-}

_____ 9. Which electron configuration represents an atom that is likely to form an ionic bond?
  (1) 2
  (2) 2-8
  (3) 2-8-8
  (4) 2-8-8-2

_____ 10. The formation of an ionic bond involves
  (1) cations, only.
  (2) anions, only.
  (3) the transfer of electrons between atoms.
  (4) noble gases.

Names and Formulas

Writing the formula for an ionic compound is simple if you keep one rule in mind: All ionic compounds are neutral. As a result, the total charge of the cations must equal the total charge of the anions.

As you have seen the formula for the salt magnesium chloride is MgS. The cation, Mg^{2+}, and the anion, S^{2-}, carry a double charge, although opposite in sign. To be neutral, this ionic com-

Unit V Ionic Compounds continued

pound must be made up of equal numbers of Mg^{2+} cations and S^{2-} anions. As a result, the formula for magnesium sulfide is written as MgS to show this one-to-one ratio. This one-to-one ratio is also seen in the formula for sodium chloride, NaCl. This salt is neutral because both ions carry a single charge.

Self-Check Write the formula for aluminum nitride, which consists of Al^{3+} and N^{3-}.

You must take care, however, when writing the formula for an ionic compound where the charges of the cation and anion differ. Consider the example of potassium oxide. The potassium cation is K^+, while the oxide anion is O^{2-}. These ions must be combined in such a way so that the compound is neutral.

To determine how to get a neutral compound, look for the lowest common multiple of the charges on the ions. The lowest common multiple of 1 and 2 is 2. Therefore, the formula should indicate two positive charges and two negative charges. For two positive charges, you need two K^+ cations. For two negative charges, you need only one O^{2-} anion. Therefore, the formula for potassium oxide is written as K_2O.

The 2 in the above formula is written as a subscript to indicate that the compound contains K^+ ions and O^{2-} ions in a two-to-one ratio. This ratio results in a neutral compound.

To write a formula, you must obviously know the charges on both the cation and anion in the ionic compound. This information can be found in a periodic table. For example, if you check a periodic table and look at the information listed for oxygen, you will see a –2. The key will tell you that this value represents oxygen's oxidation state. Oxidation states are covered in more detail in a later unit. At this point, all you need to know that the –2 represents the charge on an oxide anion.

This –2 charge makes sense if you look at oxygen's electron configuration: 2-8-6. An oxygen atom requires two additional electrons to become stable. The addition of two electrons gives the oxide anion a –2 charge.

Self-Check Write the formula for aluminum chloride.

If you check a period table, you will notice that many elements have several charges listed. For example, sulfur, S, has listed three charges: –2, +4, and +6. The first one listed, –2, is the most common charge found on a sulfide anion. Therefore, when writing a formula for an ionic compound, always use the first charge listed for an element on a periodic table unless told otherwise.

Notes/Study Ideas/Answers

Unit V Ionic Compounds continued

Notes/Study Ideas/Answers

Examples of when you select a charge other than the first one listed can be seen when writing the formulas for certain ionic compounds. Compounds that contain copper, Cu, are one example. Copper belongs to a group of metals that can become stable without achieving a noble gas configuration. A copper atom can lose either one or two electrons, depending on the other atom with which it is reacting. A copper atom that loses one electron becomes a Cu^+ cation. A copper atom that loses two electrons becomes a Cu^{2+} cation.

Obviously, both these ions cannot be called copper cations. If they were, you would have no way of telling whether the cation is Cu^+ or Cu^{2+}. To distinguish between these two cations, the Cu^+ cation is written as copper(I) cation and the Cu^{2+} is written as copper(II) cation. These names are read as the "copper one cation" and the "copper two cation", respectively. Therefore, the formula for copper(I) chloride is written as CuCl, while the formula for copper(II) chloride is written as $CuCl_2$.

Self-Check Write the formula for iron(III) oxide.

Another caution that you must exercise when writing formulas involves ionic compounds that contain **polyatomic ions**. A polyatomic ion is an ion made of two or more atoms that act as a unit. Examples of polyatomic ions include NH_4^+, OH^-, NO_3^-, and SO_4^{2-}.

Like all ionic compounds, those that contain polyatomic ions are neutral. Their formulas must reflect their neutrality. Therefore, you use the same procedure for writing formulas that involve polyatomic ions, with one addition. Parentheses are used whenever a polyatomic ion is present more than once.

For example, consider the formula for an ionic compound that consists of ammonium cations, NH_4^+, and sulfate anions, SO_4^{2-}. To be neutral, the compound must contain a 2:1 ratio of NH_4^+ cations to SO_4^{2-} anions. The formula is therefore written as $(NH_4)_2SO_4$. The parentheses show that everything inside the parentheses is doubled by the subscript 2 outside.

Self-Check Write the formula for copper(II) phosphate. Reference Table E lists Selected Polyatomic Ions, including both their formulas and charges.

Compounds that consist of only two elements are known as **binary compounds**. Examples of binary compounds include NaCl, $CaCl_2$, Al_2O_3, and K_2O. Notice that binary compounds contain only two elements, but they may contain any number of atoms.

Name _____ Class _____ Date _____

Unit V Ionic Compounds continued

In contrast, compounds that contain a polyatomic ion are not binary compounds. For example, the compound Fe(OH)$_3$ contains three elements—iron (Fe), oxygen (O), and hydrogen (H).

REVIEW YOUR UNDERSTANDING

_____ 11. The formula for any ionic compound must
   (1) contain an equal ratio of cations to anions.
   (2) contain more cations than anions, if a polyatomic ion is present
   (3) be written so that the anion is placed before the cation.
   (4) reflect the compound's neutrality.

_____ 12. Which represents the correct symbol for the iron(III) cation?
   (1) Fe^{2+}
   (2) Fe^{3+}
   (3) Fe^3
   (4) Fe^{3-}

_____ 13. Which represents the correct formula for copper(I) oxide?
   (1) CuO
   (2) Cu_2O
   (3) CuO_2
   (4) Cu_2O_2

_____ 14. Which represents the correct formula for ammonium dichromate?
   (1) $NH_4Cr_2O_7$
   (2) $NH_4(Cr_2O_7)_2$
   (3) $(NH_4)_2Cr_2O_7$
   (4) $(NH_4)_4(Cr_2O_7)_2$

_____ 15. Which compound does not contain a polyatomic ion?
   (1) sodium carbonate
   (2) sodium sulfate
   (3) sodium sulfite
   (4) sodium sulfide

Notes/Study Ideas/Answers

Properties of Ionic Compounds

An ionic bond usually forms between a metal and nonmetal to produce an ionic compound that has distinctive properties. These properties result from the fact that ionic bonds are strong.

Copyright © by Holt, Rinehart and Winston. All rights reserved.

Ionic Compounds

Focus On Regents Exam

Unit V Ionic Compounds continued

These strong bonds result in crystalline compounds that are hard and brittle. Hard means the crystal is able to resist a large force applied to it without breaking. Brittle means that when the applied force becomes too strong to resist, the crystal develops a widespread fracture rather than a small dent. The ions in a crystal are arranged in a repeating pattern, forming layers. Each layer is position so that a cation in one layer is next to an anion in the next layer.

As a result of their strong bonds, ionic compounds also have high melting and boiling points. For example, magnesium fluoride melts at 1261°C and boils at 2239°C. Compare those values to those for water, which is not an ionic compound. Ice melts at 0°C, and water boils at 100°C.

With such high boiling points, ionic compounds are rarely gaseous at room temperature. Sodium chloride, for example, will remain a solid no matter how long it remains at room temperature.

As solids, ionic compounds do not conduct electricity. However, if the salt is heated to its melting point, the liquid will conduct electricity. Molten salts are good conductors of electricity, although they do not conduct as well as metals. Also, if a salt is dissolved in water, the solution can conduct an electric current. Both molten and dissolved salts conduct electricity because their charged particles are free to move.

REVIEW YOUR UNDERSTANDING

_____ 16. Which compound contains ionic bonds?
  (1) N_2O
  (2) CO
  (3) CH_4
  (4) Na_2O

_____ 17. Identity a property of ionic compounds that exist in the solid state.
  (1) ability to conduct electricity
  (2) low boiling point
  (3) high melting point
  (4) crushes easily when hammered or pounded

_____ 18. Which compound has a high melting point?
  (1) CO_2
  (2) $MgCl_2$
  (3) CS_2
  (4) NH_3

Unit V Ionic Compounds continued

___ 19. An ionic compound conducts electricity only when
(1) its ions are free to move.
(2) it contains a polyatomic ion.
(3) the compound is dissolved in water.
(4) the compound is molten.

ANSWERS TO SELF-CHECK QUESTIONS

- $K + energy \longrightarrow K^+ + e^-$

- $Al + energy \longrightarrow Al^{3+} + 3e^-$

- $O + 2e^- \longrightarrow O^{2-} + energy$

- One atom gives up one or more electrons and becomes a cation. Another atom accepts one or more electrons and becomes an anion. An attractive force joins the cation and anion, forming an ionic bond.

- magnesium oxide

- AlN

- $AlCl_3$

- Fe_2O_3

- $Cu_3(PO_4)_2$

Notes/Study Ideas/Answers

| UNIT V | Holt Chemistry: The Physical Setting |

Questions for Regents Practice

Ionic Compounds

PART A

_____ 1. Which element forms a cation?
   (1) oxygen
   (2) neon
   (3) barium
   (4) fluorine

_____ 2. Which element forms an anion?
   (1) sodium
   (2) sulfur
   (3) strontium
   (4) potassium

_____ 3. Which element does not form either cations or anions?
   (1) silicon
   (2) copper
   (3) helium
   (4) vanadium

_____ 4. The atoms of nonmetal elements tend to form
   (1) anions that have radii larger than the atoms.
   (2) anions that have radii smaller than the atoms.
   (3) cations that have radii larger than the atoms.
   (4) cations that have radii smaller than the atoms.

_____ 5. Which is a binary compound?
   (1) O_2
   (2) NaO
   (3) $NaHCO_3$
   (4) $NaClO_3$

_____ 6. What always occurs when an ionic bond forms?
   (1) Energy is absorbed.
   (2) Energy is released.
   (3) The cations and anions combine in a 1:1 ratio.
   (4) A charged compound is formed.

_____ 7. An ionic bond forms when
   (1) electrons are transferred between two atoms.
   (2) electrons are shared between two atoms.
   (3) two atoms achieve an electron configuration similar to that of an alkali metal.
   (4) two nonmetal elements react.

_____ 8. What is the correct name for the ionic compound NH_4Cl?
   (1) nitrogen chloride
   (2) nitrogen chlorate
   (3) ammonium chloride
   (4) ammonium chlorate

_____ 9. What is the correct formula for copper(II) hydroxide?
   (1) CuOH
   (2) $Cu(OH)_2$
   (3) Cu_2OH
   (4) $CuOH_2$

_____ 10. Which is a polyatomic ion?
   (1) nitrogen
   (2) nitride
   (3) nitrite
   (4) nitrogen dioxide

Unit V Ionic Compounds continued

_____ 11. Which describes an ionic solid?
   (1) high melting point
   (2) excellent conductor
   (3) gas at room temperature
   (4) low boiling point

PART B-1

_____ 12. Which particle has the same electron configuration as a potassium ion?
   (1) neon atom
   (2) argon atom
   (3) fluoride ion
   (4) sodium ion

_____ 13. Which represents the electron configuration of a Mg^{2+} cation?
   (1) 2-8
   (2) 2-8-1
   (3) 2-8-2
   (4) 2-8-8

_____ 14. The atom of which element has an ionic radius smaller than its atomic radius?
   (1) oxygen
   (2) nitrogen
   (3) sulfur
   (4) cesium

_____ 15. An ionic compound has the formula XO. Which group on a periodic table contains element X?
   (1) Group 1
   (2) Group 2
   (3) Group 17
   (4) Group 18

_____ 16. A pure substance melts at 770°C and conducts electricity when it is either melted or dissolved in water. Which formula represents this substance?
   (1) KCl
   (2) CCl_4
   (3) HCl
   (4) SO_2

PART B-2

_____ 17. Complete the table below.

Element	Ion	Name of ion
Barium	Ba^{2+}	
Chlorine		chloride
Chromium	Cr^{3+}	
Fluorine	F^-	
Manganese		manganese(II)
Oxygen		oxide

_____ 18. Write the formula for each of the following ionic compounds.
   (1) magnesium bromide
   (2) ammonium dichromate
   (3) cobalt(III) sulfate

_____

_____

_____

Name _____ Class _____ Date _____

Unit V Ionic Compounds continued

PART C

Imagine that you are asked to write a laboratory exercise that investigates the properties of ionic compounds. Complete each of the following instructions, telling the reader what to expect if an unknown substance is an ionic compound.

19. Tap the substance gently. [Describe how to determine if the substance is an ionic compound.]

_____

_____

_____

_____

_____

_____

_____

20. Dissolve a sample of the substance in water. [Describe how to determine if the substance is an ionic compound.]

_____

_____

_____

_____

_____

_____

_____

_____

UNIT VI
Covalent Compounds

Holt Chemistry: The Physical Setting

Not all atoms achieve stability by either losing or gaining electrons. Consider carbon as an example. Carbon's atomic number is 6. A carbon atom, therefore, has six protons and six electrons. Its six electrons have the configuration 2-4. To achieve the same configuration as a noble gas, a carbon atom can lose its four valence electrons. The result would be an ion with just two electrons in its first shell, which is the same electron configuration as helium.

However, a carbon atom can just as easily gain four electrons to have the same electron configuration as neon: 2-8. Rather than either lose or gain electrons, a carbon atom forms a bond in another way. Carbon atoms share electrons with other atoms. This sharing of electrons produces a **covalent bond**. Compounds that contain covalent bonds are called **molecules**.

Covalent Bonds

Hydrogen is another element that forms a covalent bond. With only one electron in its first shell, a hydrogen atom needs one additional electron to achieve the same configuration as that of a helium atom. Like helium, a hydrogen atom with two electrons is stable.

Two hydrogen atoms represent perfect partners to form a covalent bond. Each atom has one electron. By sharing their one electron with one another, both hydrogen atoms can have two electrons at times. This sharing results in a covalent bond. The bonded atoms have a lower energy level and more stability than the individual atoms. Thus, the formation of a covalent bond releases energy.

As a rule, metals tend to react with nonmetals to form ionic bonds. Nonmetals tend to react with other nonmetals to form covalent bonds. However, some elements can form either an ionic bond or a covalent bond. The type of bond they form depends on the other atom or atoms that are present.

Consider the example of chlorine. In the presence of sodium, chlorine will form an ionic bond. However, if two chlorine atoms are present, they will form a covalent bond with one another.

A chlorine atom has seven valence electrons. If each chlorine atom were to share one of its valence electrons with another chlorine atom, then each chlorine atom can have eight valence electrons at times.

What You'll Need to Learn
This topic is part of the Regents Curriculum for the Physical Setting Exam.
Standard 4, Performance Indicator 5.2: *Explain chemical bonding in terms of the behavior of electrons.*

What Terms You'll Need to Know
covalent bond
hydrogen bond
intermolecular force
Lewis structure
metallic bond
molecule
molecular formula
nonpolar covalent bond
polar covalent bond
structural formula

Where You Can Learn Even More:
Holt Chemistry: The Physical Setting
Chapter 6: Covalent Compounds
Chapter 11: States of Matter (section 2)

Unit VI Covalent Compounds *continued*

Notes/Study Ideas/Answers

The covalent bond that forms between two atoms is represented by a dash. For example, the covalent bond formed between two hydrogen atoms and between two chlorine atoms is drawn as follows.

$$\text{H}-\text{H} \qquad \text{Cl}-\text{Cl}$$

The diagrams above are known as **structural formulas**. A structural formula shows what atoms are present in a compound and how these atoms are bonded. The structural formulas shown above indicate that the two atoms in each compound form a *single covalent bond* by sharing a pair of electrons.

Structural formulas provide more information than a **molecular formula**, which indicates only the kind and number of atoms present. For example, the molecular formulas for the compounds shown above are written as H_2 and Cl_2, respectively. Notice that a molecular formula does not reveal how the atoms are bonded.

> **Self-Check** Write the molecular formula and draw the structural formula for two iodine (I) atoms that form a covalent bond.

Some atoms cannot achieve stability by forming a covalent bond where they share a pair of electrons. Consider the example of oxygen, which is an element that can form either an ionic bond or a covalent bond. With six valence electrons, an oxygen atom requires an additional two electrons to have the same configuration as a noble gas. Accepting two electrons from a magnesium atom to form an ionic bond can do this.

An oxygen atom can also achieve stability by bonding with another oxygen atom. However, in this case, the two atoms must share four electrons, or two pairs, to obtain eight valence electrons at times. The structural formula for two covalently-bonded oxygen atoms is shown as O=O.

This structural formula shows that two oxygen atoms form a *double covalent bond*. Keep in mind that a double bond represents four electrons, or two pairs, that are shared between the atoms.

> **Self-Check** How many electrons do the atoms share in the compound with the structural formula O=C=O?

Some atoms form a *triple covalent bond* to achieve stability. An example is N_2 whose structural formula is shown as N≡N.

A triple covalent bond involves six electrons, or three pairs, that are shared between two atoms.

Copyright © by Holt, Rinehart and Winston. All rights reserved.

Covalent Compounds — Focus On Regents Exam

Unit VI Covalent Compounds continued

Self-Check How many electrons do the atoms share in the compound with the structural formula H—C≡N?

In a covalent bond, the two atoms may share the electrons equally. This equal sharing is the result of both atoms having the same attractive force for the shared electrons. In the case of O_2, for example, each oxygen atom has eight protons in its nucleus. The eight protons in each nucleus have the same attractive force for the shared pair of electrons that form the covalent bond. When electrons are shared equally between two atoms, the bond is called a **nonpolar covalent bond**.

In some covalent bonds, one of the atoms has a greater attractive force for the shared electrons. The covalent bond that forms between a hydrogen atom and a chlorine atom is an example. These two atoms share a pair of electrons. As a result, the hydrogen atom achieves stability by having two valence electrons in its first shell at times. The chlorine atom achieves stability by having eight valence electrons at times. The following structural formula shows the covalent bond between hydrogen and chlorine.

H—Cl

However, the sharing of the electrons is not equal. A hydrogen atom has only one proton, while a chlorine atom has 17 protons. As a result, a chlorine atom can attract the shared pair of electrons with a greater force. The shared electrons spend more time nearer to the chlorine atom. When two atoms share electrons unequally, the bond is called a **polar covalent bond**.

Self-Check Are the covalent bonds in O=C=O polar or nonpolar? Explain your answer.

Atoms that are covalently bonded can be shown with the use of **Lewis structures**. A Lewis structure is a structural formula in which the valence electrons are represented by dots and the covalent bonds by dashes.

When drawing a Lewis structure, you represent the nucleus and all the electrons, except the valence electrons, by the element's symbol. Each valence electron is represented by a dot. Therefore, the Lewis structure for hydrogen is drawn as follows.

H·

Unit VI Covalent Compounds *continued*

Notes/Study Ideas/Answers

The Lewis structure of a carbon atom with its four valence electrons is drawn as follows.

$$\cdot \overset{\cdot}{\underset{\cdot}{C}} \cdot$$

Notice that when the dots for the valence electrons are placed around the symbol, each side must contain an unpaired electron before any side can contain a pair of electrons.

Self-Check Draw the Lewis structure of a chlorine atom.

Lewis structures can also be drawn for atoms that are covalently bonded. For example, the Lewis structure for HCl is drawn as follows.

$$H\cdot \; + \; \cdot\ddot{\underset{\cdot\cdot}{Cl}}: \; \longrightarrow \; H-\ddot{\underset{\cdot\cdot}{Cl}}:$$

Notice that the shared pair of electrons is represented by a dash while the each unshared electron is represented by a dot.

Self-Check Draw the Lewis structure for Cl_2.

Lewis structures can also be drawn for molecules that contain double and triple covalent bonds. For example, the Lewis structure for N_2 is drawn as follows.

$$:N::N: \; \text{or} \; :N\equiv N:$$

Ionic compounds can also be illustrated with the use of Lewis structures. An example is calcium chloride, which is illustrated with the following Lewis structures.

$$[Ca]^{2+} \; \text{and} \; [:\ddot{\underset{\cdot\cdot}{Cl}}:]^- \; \text{and} \; [:\ddot{\underset{\cdot\cdot}{Cl}}:]^-$$

$$[Ca]^{2+} \; \text{and} \; 2[:\ddot{\underset{\cdot\cdot}{Cl}}:]^-$$

REVIEW YOUR UNDERSTANDING

_____ 1. Which two elements are likely to form a covalent bond?
 (1) K and O
 (2) N and S
 (3) H and Ne
 (4) Ca and Cl

Unit VI Covalent Compounds continued

_____ 2. Which structural formula includes a nonpolar covalent bond?
   (1) H—S—H
   (2) H—Br
   (3) Cl—O—Cl
   (4) Br—Br

_____ 3. In a polar covalent bond, the electrons are
   (1) shared equally between the atoms.
   (2) shared unequally between the atoms.
   (3) transferred from one atom to another atom.
   (4) removed from both atoms.

_____ 4. Which combination of elements can form a polar covalent bond?
   (1) H and Br
   (2) N and N
   (3) Na and Br
   (4) Na and N

_____ 5. Which Lewis electron-dot structure is drawn correctly for the atom it represents?
   (1) :Ṅ
   (2) :F̈:
   (3) :Ö:
   (4) ·Ċ·

Electronegativity Values

Electronegativity values can be used to determine the type of bond that two elements will form. Recall that an electronegativity value represents the ability of an atom in a compound to attract the bonding electrons. Electronegativity values have been set up as an arbitrary scale.

Because it has the greatest attraction for bonding electrons, a fluorine atom was assigned the highest electronegativity value. The electronegativity value of fluorine is 4.0. All other atoms were assigned values based on how their attraction for bonding electrons compared to that of fluorine's. You can locate the electronegativity value of an element by checking *Reference Table S, Properties of Selected Elements*.

Self-Check What are the electronegativity values for sodium, silver, and neon?

Unit VI Covalent Compounds continued

Obviously, if the two atoms that form a covalent bond are the same element, then the difference between their electronegativity values is 0. For example, both atoms in F_2 have an electronegativity value of 4.0, resulting in a difference of 0. When the difference between the electronegativity values of two atoms is 0, then the atoms form a nonpolar covalent bond.

In fact, if the difference in electronegativity values falls between 0 and 0.5, then the atoms form a nonpolar covalent bond. For example, carbon has an electronegativity value of 2.6 and nitrogen has an electronegativity value of 3.0. The difference between these two values is 0.4, meaning that carbon and nitrogen form a nonpolar covalent bond.

Self-Check Do nitrogen and oxygen atoms form a nonpolar covalent bond? Explain your answer.

A difference in electronegativity values that falls between 0.5 and 2.1 results in the formation of a polar covalent bond. The closer the difference is to 2.1, the greater the polarity of the bond. For example, the difference between the electronegativity value of hydrogen (2.1) and fluorine (4.0) is 1.9. The difference between the electronegativity value of oxygen (3.4) and fluorine (4.0) is 0.6. Both combinations of elements form polar covalent bonds. However, the bond between H and F has a higher degree of polarity than the bond between O and F.

Self-Check Which pair of elements forms a covalent bond that has the greater polarity: C and Cl or Si and Cl? Explain your answer.

Be aware that a molecule that has polar covalent bonds may be nonpolar. For example, examine the following structural formulas.

$$\ddot{\underset{..}{O}}=C=\ddot{\underset{..}{O}} \qquad H-\underset{\underset{H}{|}}{\overset{\overset{H}{|}}{C}}-H$$

The bonds between each pair of atoms in both these molecules are polar covalent bonds. However, both molecules are symmetrical, resulting in each molecule having a balanced charge distribution. Both molecules, therefore, are nonpolar.

Self-Check Use Lewis structures to explain why $SiCl_4$ is a nonpolar molecule, even though it contains polar bonds.

Unit VI Covalent Compounds *continued*

A difference in electronegativity values that is greater than 2.1 results in the formation of an ionic bond. For example, the electronegativity difference between magnesium and oxygen is great enough for an O atom to remove two electrons from an Mg atom to form an ionic bond. Keep in mind that the boundaries between bond types are arbitrary. Using the difference in electronegativity values is just one way to classify bonds.

Another way to classify bonds is by examining the properties of the compound. You learned that ionic compounds have high melting points and boiling points. Ionic compounds are also hard and brittle. They do not conduct electricity unless they are molten or dissolved.

In contrast, covalent compounds, or molecules, have much lower melting and boiling points than ionic compounds. As a result, many covalent compounds exist as gases at room temperature. Covalent compounds do not conduct electricity, even when they are molten or dissolved. They do not conduct electricity because covalent compounds, unlike ionic compounds, do not consist of charged particles.

Self-Check Prepare a table summarizing the differences between ionic solids and covalent compounds with respect to their physical properties, including hardness, melting and boiling points, and ability to conduct electricity.

Some compounds contain both covalent and ionic bonds. Two examples are sodium nitrate, $NaNO_3$, and potassium sulfate, K_2SO_4. Notice that both these compounds contain a polyatomic ion. The atoms that make up a polyatomic ion are covalently bonded to one another. For example, examine the following Lewis structure for sulfate.

$$:\ddot{O}:$$
$$:\ddot{O}:\ddot{S}:\ddot{O}:$$
$$:\ddot{O}:$$

Notice that each of the four O atoms shares a pair of electrons with the central S atom. Each dash in the Lewis structure of a sulfate polyatomic ion represents a polar covalent bond.

However, polyatomic ions, such as sulfate, combine with other ions by forming ionic bonds. For example, a sulfate anion, with its 2– charge, forms an ionic bond with two potassium cations, each with a 1+ charge, in the ionic compound K_2SO_4.

Notes/Study Ideas/Answers

Unit VI Covalent Compounds *continued*

REVIEW YOUR UNDERSTANDING

_____ 6. The atoms of which element have the weakest attractions for bonding electrons?
(1) P
(2) Si
(3) Na
(4) S

_____ 7. In contrast to ionic compounds, covalent compounds
(1) have higher melting points and boiling points.
(2) conduct electricity in the solid state.
(3) are hard and brittle.
(4) can exist as gases at room temperature.

_____ 8. In which compound does the bond between the atoms have the *least* polar character?
(1) H_2S
(2) H_2O
(3) HF
(4) HI

_____ 9. Which compound has the greatest degree of ionic character?
(1) SiF_4
(2) AlF_3
(3) MgF_2
(4) NaF

_____ 10. Identify the molecule that is nonpolar and contains a nonpolar covalent bond.
(1) H_2O
(2) CH_4
(3) I_2
(4) CO_2

Other Bond Types

In addition to ionic and covalent bonding, atoms can bond in another way. This type of bond is known as a **metallic bond**. Obviously, this type of bonding is found in metals such as copper, magnesium, and gold. Metallic bonds are the result of the attraction between the valence electrons of each metal atom and all the other atoms in the metal.

The many billions of atoms that make up metal, such as a piece of aluminum foil, are held in the solid because all the valence electrons are attracted to all the atoms in the solid. In effect, the valence electrons do not belong to a single atom. Rather they belong to all the atoms in the metal.

You learned that metals are excellent conductors of electricity. Metallic bonding is what accounts for this physical property. The nucleus and inner electrons of a metal atom remains in a fixed position. However, the valence electrons of the metal atom are free to roam around the solid. You can think of them as a "sea" of mobile electrons that are free to move throughout the metal. As a result, an electric current easily passes through a metal.

Self-Check Explain why Hg conducts electricity even though it is a liquid at room temperature.

You learned that even a tiny ionic crystal consists of many billions of ions. Each cation in the crystal attracts more than one anion. In turn, each anion attracts more than one cation. All these attractive forces result in a tightly packed structure, or salt crystal.

Covalent molecules also attract one another, although not nearly as strongly as do the ions in a crystal. The attractive forces that act between molecules are called **intermolecular forces**. The strength of intermolecular forces varies, depending on the nature of the molecule.

An example of a strong intermolecular force is known as a **hydrogen bond**. A hydrogen bond forms when a hydrogen atom that is bonded to a highly electronegative atom of one molecule is attracted to two unshared electrons of another molecule. A hydrogen bond forms between a hydrogen atom on one molecule and usually an atom of fluorine, oxygen, or nitrogen on another molecule.

Water is a perfect example of how hydrogen bonds affect the properties of the substance. To understand how a hydrogen bond forms, first examine the Lewis structure of a water molecule.

$$\overset{O}{\underset{H \quad H}{\wedge}}$$

The bond between the O atom and each H atom is a polar covalent bond. As a result, each H atom has a large, partially positive charge. In contrast, the O atom has a large, partially negative charge. Therefore, water is a polar molecule with positive and negative regions.

Self-Check How can you determine that the bonds in a water molecule are polar covalent bonds?

Notes/Study Ideas/Answers

Unit VI Covalent Compounds continued

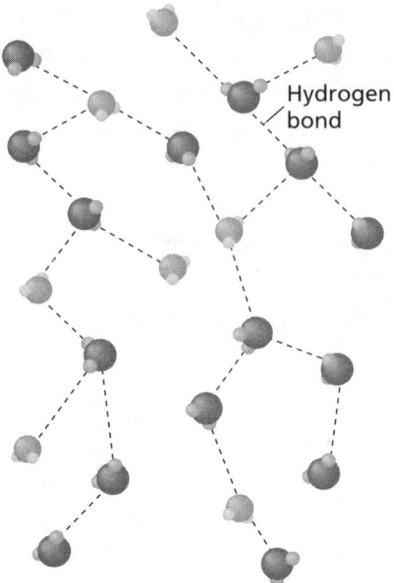

Liquid water

Notice that the O atom in a water molecule has two unshared pairs of electrons. The partially positive hydrogen atom on a water molecule can be attracted to the unshared pairs of electrons on the oxygen atom of a neighboring water molecule. This attraction creates a hydrogen bond, as shown in the following illustration.

A hydrogen bond is a weak bond. However, because there are so many hydrogen bonds formed between the many billions of water molecules in even a tiny drop, their combined force of attraction is rather strong. This strong attractive force accounts for the high boiling point of water compared to molecules that have a similar composition, such as H_2S.

Intermolecular forces also exist between covalent compounds that do not form hydrogen bonds. For example, intermolecular forces weakly attract HBr molecules to one another. These attractions are the result of HBr being a polar molecule with positive and negative regions. However, the force of attraction between HBr molecules is not as strong as it is between HF molecules, which are attracted to one another by hydrogen bonds.

Unit VI Covalent Compounds continued

REVIEW YOUR UNDERSTANDING

_____ 11. Which contains metallic bonds?
(1) F_2
(2) Fe
(3) FeO
(4) $FeSO_4$

_____ 12. Which molecules can attract one another by forming hydrogen bonds?
(1) H_2
(2) H_2SO_4
(3) H_2CO_3
(4) HF

_____ 13. A solid substance is an excellent conductor of electricity. The bonding in this substance is
(1) nonpolar covalent.
(2) polar covalent.
(3) ionic.
(4) metallic.

_____ 14. In which type of bonding are the valence electrons free to move throughout the substance?
(1) ionic
(2) covalent
(3) metallic
(4) hydrogen

ANSWERS TO SELF-CHECK QUESTIONS

- molecular formula: I_2; structural formula I—I

- The C atom shares four electrons, or two pairs, with each O atom.

- The H and C atoms share a pair of electrons, while the C and N atoms share six, or three pairs of electrons.

- The bonds are polar because each O atom, with its eight protons, attracts the shared pairs of electrons more than does the C atom with its six protons.

- :C̈l·

- :C̈l:C̈l:

Unit VI Covalent Compounds *continued*

- Table S lists the following electronegativity values: Na: 0.9; Ag: 1.9; Ne: no value listed because atoms of the noble gases have no attraction for electrons

- The difference in electronegativity values between N and O is 0.4, resulting in these two atoms forming a nonpolar covalent bond.

- The difference in electronegativity values between C and Cl is 0.6, while the difference between Si and Cl is 1.3. As a result, the covalent bond that forms between Si and Cl has a greater polarity.

- The Lewis structure for $SiCl_4$ shows that the molecule is symmetrical and therefore is nonpolar.

$$\begin{array}{c} Cl \\ | \\ Cl-Si-Cl \\ | \\ Cl \end{array}$$

Property	Ionic Compound	Covalent Compound
hardness	hard	soft
melting point	high	low
boiling point	high	low
electrical conductivity	when melted or dissolved	none

- Mercury (Hg) is a metal and therefore an excellent conductor of electricity because of the presence of metallic bonds.

- Calculate the difference in the electronegativity values for H and O.

Covalent Compounds

PART A

1. When a covalent bond is formed,
   (1) the bonded atoms have less stability.
   (2) electrons are transferred between the bonded atoms.
   (3) energy is absorbed.
   (4) energy is released.

2. Which formula represents a molecule?
   (1) Ar
   (2) $NaNO_3$
   (3) $C_6H_{12}O_6$
   (4) CaO

3. Which of the following atoms has the greatest tendency to attract electrons?
   (1) boron
   (2) bromine
   (3) barium
   (4) beryllium

4. Electronegativity is defined as an atom's
   (1) ability to lose electrons and form an ionic bond.
   (2) ability to attract the bonding electrons in a compound.
   (3) tendency to form the electron configuration of a noble gas.
   (4) ability to have its valence electrons move freely throughout the substance.

5. Which element has the highest electronegativity?
   (1) Ca
   (2) K
   (3) H
   (4) Al

6. The strongest forces of attraction occur between molecules of
   (1) HF
   (2) HI
   (3) HCl
   (4) He

7. What is represented by the dots in a Lewis electron-dot diagram of an atom of an element in Group 18 of a periodic table?
   (1) the number of neutrons in the atom
   (2) the number of protons and neutrons in the atom
   (3) the total number of electrons in the atom
   (4) the number of valence electrons in the atom

8. Which formula represents a substance that has a low melting point and does not conduct electricity when dissolved in water?
   (1) Na_2SO_4
   (2) CS_2
   (3) NH_4Cl
   (4) KCl

Unit VI Covalent Compounds continued

PART B-1

____ 9. What type of bond forms between a hydrogen atom and an iodine atom?
(1) ionic bond
(2) nonpolar covalent bond
(3) polar covalent bond
(4) hydrogen bond

____ 10. The table below represents the normal boiling point of four compounds.

Compound	Boiling Point
HCl	−83.7°C
CH_3F	−78.6°C
CH_3Cl	−24.2°C
HF	19.4°C

Which compound has the strongest intermolecular forces?
(1) HCl
(2) CH_3F
(3) CH_2Cl
(4) HF

____ 11. A student performs the same test on two white crystalline solids, A and B. The results are shown below.

	Solid A	Solid B
Melting point	862°C	125°C
Electrical conductivity when dissolved	good	none

The results of these tests suggest that
(1) both solids contain only ionic bonds.
(2) both solids contain only covalent bonds.
(3) Solid A contains only covalent bonds and solid B contains only ionic bonds.
(4) Solid A contains only ionic bonds and solid B contains only covalent bonds.

____ 12. Which of the following solids has the lowest melting point?
(1) Na_2O
(2) H_2O
(3) MgO
(4) $CaPO_4$

PART B-2

Each molecule listed below is formed by covalent bonds.
Molecule A: F_2
Molecule B: CCl_4
Molecule C: NH_3

13. Explain why CCl_4 is classified as a nonpolar molecule.

_____

_____

_____

14. Draw the Lewis structure for the NH_3 molecule.

_____

_____

_____

15. Explain why NH_3 has stronger intermolecular forces of attraction than F_2.

_____

_____

_____

16. Explain why the melting and boiling points of covalent compounds are usually lower than those of ionic compounds.

_____

_____

_____

Unit VI Covalent Compounds continued

17. Both ammonia, NH_3, and methane, CH_4, are covalent compounds. Yet the boiling point of ammonia is 130°C higher than that of methane. What might account for this difference in their boiling points?

PART C

Base your answers to questions 18 through 20 on the article below and on your knowledge of chemistry.

Garbage

The largest structure ever made by humans is the Fresh Kills landfill in Staten Island, New York. Although a landfill is not a structure in the true meaning of the word, Fresh Kills does represent the largest collection of materials ever accumulated for a single purpose by humans. Fresh Kills began in 1948 atop a swamp. For over 50 years, the landfill received nearly 13,000 tons of garbage every day until it was closed in 2003. If all the garbage in Fresh Kills were stacked on a 1000 football fields, it would pile up as high as a 21-story building.

Much of the garbage was dumped into Fresh Kills in trash bags. These bags are made from a polyethylene, a molecule that does not break down easily. Polyethylene consists of carbon and hydrogen atoms bonded together to form long chains, often as long as 7000 atoms bonded to one another.

18. Draw a Lewis structure to show how the two atoms in a polyethylene molecule bond to one another.

Unit VI Covalent Compounds continued

19. Describe the type of bond that forms between these two atoms.

20. Trash bags made of polyethylene do not degrade very easily. What does this tell you about the bonds in a polyethylene molecule?

Name _____ Class _____ Date _____

UNIT VII
Chemical Reactions

Holt Chemistry: The Physical Setting

You learned that a chemical change involves a change in the identity of a substance. You also learned that a chemical reaction is the process by which a substance undergoes a chemical change. A chemical reaction is a process that creates one or more new substances whose chemical and physical properties differ from those of the original substances. Recall that the original substances are called reactants, while the substances created are called products.

Energy Changes

Whenever a chemical reaction occurs, an energy change also occurs. The energy change can be exothermic or endothermic. For example, the following chemical reaction is an example of an exothermic reaction.

methane + oxygen ⟶ carbon dioxide + water + energy

Notice that in an exothermic reaction, energy can be considered a product of the reaction and included on the right side of the arrow.

Now examine the following chemical reaction.

potassium chlorate + energy ⟶ potassium chloride + oxygen

This is an example of an endothermic reaction in which energy is considered a reactant and included on the left side of the arrow.

Although energy may be released or absorbed, energy is neither created nor destroyed. In a chemical reaction, energy is only changed from one form to another. For example, all compounds contain a form of energy known as **chemical energy**. This chemical energy is present in the bonds that hold the atoms together.

In a chemical reaction, energy is required to break a bond in a reactant. The atoms can then rearrange themselves and form new substances. When the atoms bond to form a product in a chemical reaction, energy is released.

The quantity of chemical energy present in the reactants may be greater than the quantity of energy present in the products. In this case, the chemical reaction is exothermic.

The energy changes that occur in an exothermic reaction can be plotted. Examine the following graph that shows how energy changes as an exothermic reaction progresses. The

What You'll Need to Learn
This topic is part of the Regents Curriculum for the Physical Setting Exam.
Standard 4, Performance Indicator 3.2: *Use atomic and molecular models to explain common chemical reactions.*
Standard 4, Performance Indicator 3.3: *Apply the principle of conservation of mass to chemical reactions.*
Standard 4, Performance Indicator 3.4: *Use kinetic molecular theory (KMT) to explain rates of reactions and the relationships among temperature, pressure, and volume of a substance.*
Standard 4, Performance Indicator 4.1: *Observe and describe transmission of various forms of energy.*

What Terms You'll Need to Know
activated complex
activation energy
catalyst
chemical energy
chemical equation
decomposition reaction
double replacement reaction
enzyme
law of conservation of energy

Copyright © by Holt, Rinehart and Winston. All rights reserved.

Chemical Reactions

Unit VII Chemical Reactions continued

Which Terms You'll Need to Know
law of conservation of mass
synthesis reaction
water of hydration

Where You Can Learn Even More:
Holt Chemistry: The Physical Setting
Chapter 8: Chemical Equations and Reactions
Chapter 16: Reaction Rates

exothermic reaction shown in this graph involves breaking apart HI, the reactant, to form H_2 and I_2, the products. This reaction can be written as follows.

$$\text{hydrogen iodide} \longrightarrow \text{hydrogen} + \text{iodine} + \text{energy}$$

Activation Energies for the Decomposition of HI

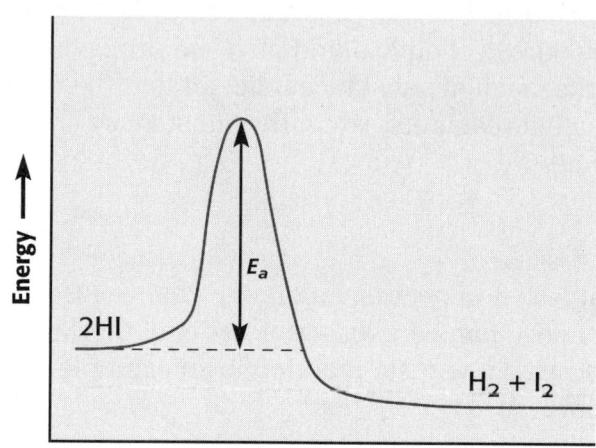

To begin the reaction, energy must be supplied to break the covalent bond between H and I in HI. The minimum energy that must be supplied to start a chemical reaction is called the **activation energy**, which is represented on the graph as the value E_a.

At the peak of its E_a value, the reactants form an **activated complex**. An activated complex is an intermediate, unstable state in a chemical reaction. An activated complex quickly breaks apart, releasing energy.

Self-Check Identify where the activated complex is located on the graph shown above.

Notice in the graph that when the reaction is complete, the products, H_2 and I_2, possess less energy than the reactant, HI. This decrease in energy does not mean that energy was destroyed. Rather, the difference in the energy possessed by the reactant and products represents the energy that was released as heat to the surroundings.

Assume that the energy released as heat is trapped and measured. Next assume that this quantity of energy as heat is added to the quantity of chemical energy possessed by the products. You would find that the total quantity of energy on the right side of the equation would equal the total quantity of energy on the left side of the equation.

Unit VII Chemical Reactions *continued*

This finding would support the **law of conservation of energy**. This law states that energy cannot be created or destroyed but can be changed from one form to another.

The law of conservation of energy also applies to endothermic reactions. Examine the following graph that shows the energy changes when HBr breaks apart to form H_2 and Br_2 in an endothermic reaction. This reaction can be written as follows.

hydrogen bromide + energy ⟶ hydrogen + bromine

Activation Energies for the Decomposition of HBr

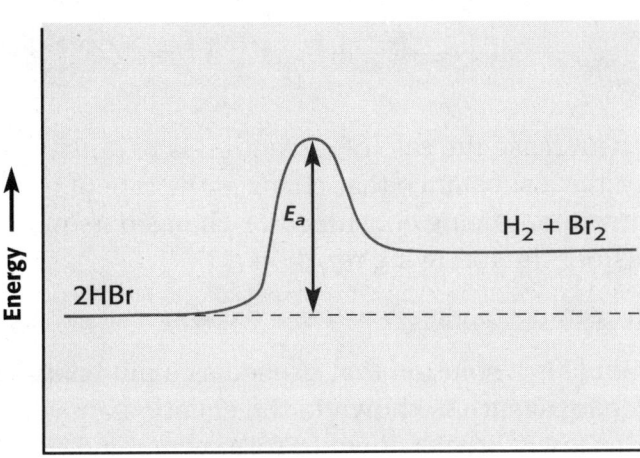

Notice that activation energy is again required for this reaction to occur. However, notice a difference from the HI graph that you examined earlier. In the graph shown above, the products, H_2 and Br_2, have more energy than the reactant, HBr. This additional energy was not created. Rather, the additional energy present in the products came from the activation energy supplied as heat.

Self-Check Sketch a graph that shows how energy changes during the following exothermic reaction: hydrogen + oxygen ⟶ water

Both exothermic and endothermic reactions will not occur unless the reactants collide under certain conditions. First, they must collide with enough energy to overcome the activation energy barrier. Second, they must collide with the proper orientation. This means that the atoms that will be rearranged must collide with one another.

If the reactants collide with the necessary energy and the proper orientation, then the reaction can proceed. Certain factors, however, can increase the rate of the reaction.

Unit VII Chemical Reactions *continued*

Increasing the concentration of the reactants increases the rate of a reaction because the reactants are more likely to collide.

Increasing the surface area also increases the rate of a reaction because more reactants are likely to collide. For example, you get a bigger blaze when you use a match to light several, small pieces of wood as compared to one, larger piece of wood.

Increasing the temperature also increases the rate of a reaction. The energy added as heat increases the chances that the reactants will collide with enough energy to overcome the activation energy barrier.

Self-Check List three ways of increasing the rate of a chemical reaction.

Another way to increase the rate of a reaction is to add a **catalyst**. A catalyst is a substance that changes the rate of a chemical reaction without being consumed or changed itself. For example, consider the following reaction.

$$\text{hydrogen peroxide} \longrightarrow \text{water} + \text{oxygen}$$

Hydrogen peroxide is a solution that is used as a mild antiseptic and a bleaching agent. As shown in the equation above, hydrogen peroxide slowly breaks down to form water and oxygen. Light energy speeds up this reaction. Therefore, hydrogen peroxide is usually stored in an opaque bottle. Even so, it still reacts slowly to form water and oxygen. If a few crystals of manganese dioxide, MnO_2, are added to hydrogen peroxide, the reaction will proceed quickly and violent. The manganese dioxide is a catalyst.

Self-Check What is a catalyst?

Catalysts function by lowering the activation energy that is required for a reaction to occur. In some cases, several difference substances may act as a catalyst for the same reaction. All the catalysts will speed up the rate of the reaction. However, some catalysts may speed up the reaction rate more than others. The catalyst that speeds up the reaction rate the most is the one that lowers the activation energy the most.

For example, let's take another look at the reaction where hydrogen peroxide breaks down into water and oxygen. In addition to MnO_2, two other catalysts can be used. These include iodine crystals and catalase, an **enzyme**. An enzyme is a substance that speeds up reaction rates in plants and animals.

Unit VII Chemical Reactions continued

The effect of each of these catalysts on hydrogen peroxide is illustrated in the following graph.

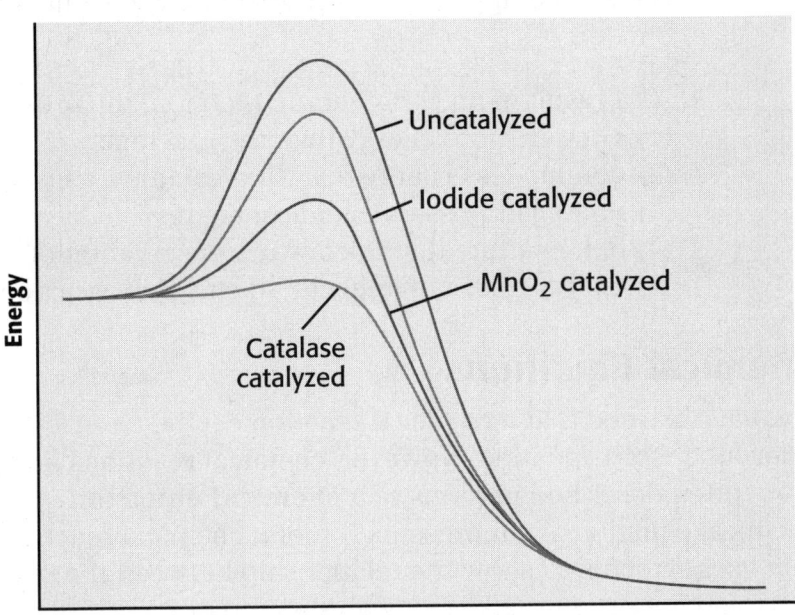

Comparison of Pathways for the Decomposition of H_2O_2

Notice that catalase is most effective at lowering the activation energy. Therefore, catalase speeds up the reaction to a greater extent than either iodine crystals or MnO_2.

REVIEW YOUR UNDERSTANDING

_____ 1. Which factor will increase the rate of a chemical reaction?
   (1) decreasing the temperature
   (2) decreasing the concentration of the reactants
   (3) increasing the energy added as heat to the reaction mixture
   (4) increasing the number of bonds present in the reactants

_____ 2. An activated complex represents the
   (1) products of a chemical reaction.
   (2) reactants of a chemical reaction.
   (3) highest energy level attained during a chemical reaction.
   (4) quantity of energy released in an exothermic reaction.

Unit VII Chemical Reactions continued

_____ 3. A catalyst
(1) raises the activation energy for a reaction.
(2) lowers the activation energy for a reaction.
(3) lowers the chances that the reactants will collide in the proper fashion for a reaction to occur.
(4) is used up during the course of a chemical reaction.

_____ 4. As a result of an endothermic reaction, the products
(1) contain more energy than the reactants.
(2) contain less energy than the reactants.
(3) no longer possess chemical energy.
(4) demonstrate that the law of conservation of energy does not apply to all chemical reactions.

Chemical Equations

You have learned that a chemical reaction can be described in terms of a word equation. However, chemical reactions are more often described in terms of a **chemical equation**. A chemical equation is a representation of a chemical reaction that uses symbols to show the relationship between the reactants and the products. A chemical equation differs from a word equation in several ways.

First, a chemical equation uses formulas for each substance involved in a reaction rather than words. For example, notice how the following word equation is rewritten as a chemical equation.

word equation: iron(III) oxide + hydrogen ⟶ iron + water

chemical equation: $Fe_2O_3 + H_2 \longrightarrow Fe + H_2O$

Self-Check Rewrite the following word equation as a chemical equation.
aluminum + copper(II) chloride ⟶
aluminum chloride + copper

In addition to formulas, chemical equations always include symbols. Two symbols that you have seen used are + and ⟶. The ⟶ symbol separates the reactants from the products. The + symbol separates each reactant and each product.

Chemical equations can also include symbols that show the physical state of each substance. An example can be seen in the following equation.

$$Na(s) + H_2O(l) \longrightarrow NaOH(aq) + H_2(g)$$

Unit VII Chemical Reactions continued

This equation tells you that Na is a solid (s), that H_2O is a liquid (l), that NaOH is dissolved in a water solution (aq) and that H_2 is a gas (g).

Self-Check Rewrite the chemical equation you wrote to include all symbols. Both aluminum and copper are solids, while both copper(II) chloride and aluminum chloride are dissolved in solution.

The following table summarizes the symbols that are often used in a chemical equation.

Symbol	Meaning
$(s), (l), (g)$	substance in the solid, liquid, or gaseous state
(aq)	substance in aqueous solution (dissolved in water)
$\rightarrow$	"produces" or "yields," indicating result of reaction
$\rightleftarrows$	reversible reaction in which products can reform into reactants; final result is a mixture of products and reactants
$\xrightarrow{\Delta}$ or $\xrightarrow{heat}$	reactants are heated; temperature is not specified
$\xrightarrow{Pd}$	name or chemical formula of a catalyst, added to speed a reaction

Chemical equations are also balanced so that they conform to the **law of conservation of mass**. This law states that in an ordinary chemical or physical change, mass is neither created nor destroyed. If you collect all the products of a reaction, you will find that their total mass is the same as the total mass of the reactants.

You learned that a chemical reaction involves a rearrangement of the atoms in the reactants. If the atoms are only rearranged, then the total number of atoms in the reactants must equal the total number of atoms in the products. A balanced chemical equation, therefore, has the same number of each type of atom on both sides of the equation. For example, consider the following equation, which shows how hydrogen gas and oxygen gas combine to form water.

$$H_2(g) + O_2(g) \rightarrow H_2O(l)$$

The reactants consist of 2 H atoms and 2 O atoms. The product consists of 2 H atoms, but only 1 O atom.

Self-Check How does the equation shown above violate the law of conservation of mass?

Unit VII Chemical Reactions continued

When balancing an equation, you cannot change the formulas. For example, you cannot balance the equation shown above by adding a 2 as a subscript in the formula for water.

$$H_2(g) + O_2(g) \rightarrow H_2O_2(l)$$

Although the above equation is balanced, the product is no longer water, but hydrogen peroxide. Therefore, adding the subscript results in a chemical reaction that does not happen.

To balance an equation, you place a coefficient in front of a formula. The coefficient multiplies the number of atoms of each element in the formula that follows. Inserting the coefficients is often a process of trial-and-error. The *Skills Toolkit* on page 268 of *Holt Chemistry* explains in detail how to balance a chemical equation.

The balanced equation for the reaction shown at the top of this page is written as follows.

$$2H_2(g) + O_2(g) \rightarrow 2H_2O(l)$$

Notice that the number of each type of atom in the reactants equals the number of each type of atom in the product. There are 4 H atoms and 2 O atoms on the left side of the equation, and there are 4 H atoms and 2 O atoms on the right side of the equation.

Self-Check Balance the following equation. $Si + CO_2 \rightarrow SiC + SiO_2$

You may find that when you balance one type of atom, another type becomes unbalanced. You may have to erase a coefficient you inserted and replace it with another one. Consider the following equation, which is partially balanced.

$$2NH_3 + O_2 \rightarrow 2NO + 3H_2O$$

Notice that there are 2 N atoms and 6 H atoms on each side of the equation. However, there are 2 O atoms on the left side and 5 O atoms on the right side. In equations where all the atoms are balanced except one that has an odd number on one side and an even number on the other side, multiply all the coefficients by 2. In this case, the equation is rewritten as follows.

$$4NH_3 + 2O_2 \rightarrow 4NO + 6H_2O$$

Unit VII Chemical Reactions continued

Both the N atoms and H atoms are balanced. However, there are 4 O atoms on the left, and 10 O atoms on the right. The coefficient for O2 must be changed.

$$4NH_3 + 5O_2 \longrightarrow 4NO + 6H_2O$$

Self-Check How many atoms of each element are present on both sides of the equation shown above?

Three additional points must be kept in mind when balancing a chemical equation. First, balance polyatomic ions as a unit when they appear on both sides of an equation. Consider the following equation.

$$AgNO_3 + AlCl_3 \longrightarrow AgCl + Al(NO_3)_3$$

The equation shown above includes the polyatomic ion nitrate, NO_3, in a reactant and in a product. Rather than balancing the N atoms and O atoms individually, treat the polyatomic ion as a unit. In this case, there is one nitrate ion on the left, and there are three nitrate ions on the right. The following equation shows how the nitrate ions are balanced as a unit.

$$3AgNO_3 + AlCl_3 \longrightarrow AgCl + Al(NO_3)_3$$

There are now three nitrate ions on both sides of the equation. However, not all the atoms have been completely balanced.

Self-Check Complete the balancing of the equation shown above.

The second point to keep in mind when balancing an equation involves **water of hydration**. Many salts contain water molecules as part of their structure. For example, the formula for iron(III) chloride is $FeCl_3 \cdot 6H_2O$. This formula tells you that six water molecules surround each $FeCl_3$. If this salt appears in a chemical equation, you must keep in mind that there are 12 H atoms and 6 O atoms in addition to the 1 Fe atom and 3 Cl atoms.

The final point to remember is that the coefficients in a balanced equation must be the smallest whole numbers. For example, the following equation is correctly balanced.

$$4KI + 2Pb(NO_3)_2 \longrightarrow 2PbI_2 + 4KNO_3$$

However, the coefficients in the above equation are not the smallest whole numbers. Therefore, the equation must be rewritten as follows.

$$2KI + Pb(NO_3)_2 \longrightarrow PbI_2 + 2KNO_3$$

Unit VII Chemical Reactions continued

REVIEW YOUR UNDERSTANDING

_____ 5. A balanced chemical equation must conform to
  (1) the law of conservation of energy, only.
  (2) the law of conservation of mass, only.
  (3) the law of conservation of energy and the law of conservation of mass, both.
  (4) neither the law of conservation of energy nor the law of conservation of mass.

_____ 6. When the equation

$$H_2 + Fe_3O_4 \rightarrow Fe + H_2O$$

is correctly balanced using smallest whole numbers, the coefficient of H_2 is
  (1) 1.
  (2) 2.
  (3) 3.
  (4) 4.

_____ 7. Which equation illustrates the law of conservation of mass?
  (1) $H_2 + I_2 \rightarrow 2HI$
  (2) $H_2 + I_2 \rightarrow HI$
  (3) $H_2 + O_2 \rightarrow 2H_2O$
  (4) $H_2 + O_2 \rightarrow H_2O$

_____ 8. Identify the equation that is correctly balanced.
  (1) $C_8H_{18} + 8O_2 \rightarrow 8CO_2 + H_2O$
  (2) $2Cu + HNO_3 \rightarrow 2Cu(NO_3)_2 + NO + H_2O$
  (3) $2Ni + 2Pb(NO_3)_2 \rightarrow 2Ni(NO_3)_2 + Pb$
  (4) $Ba(ClO_3)_2 \rightarrow BaCl_2 + 3O_2$

_____ 9. When the equation

$$Al_2(SO_4)_3 + ZnCl_2 \rightarrow AlCl_3 + ZnSO_4$$

is correctly balanced using smallest whole numbers, the sum of the coefficients is
  (1) 4.
  (2) 5.
  (3) 8.
  (4) 9.

Unit VII Chemical Reactions continued

Types of Chemical Reactions

With so many chemical reactions, grouping those that have something in common makes it easier to study them. You also have an idea of what to expect by simply looking at the reactants. There are four major reaction types that will be discussed here.

A **synthesis reaction** is the first type. A synthesis reaction is a reaction in which two or more substances combine to form a new compound. A synthesis reaction can be as simple as two elements combining to form a binary compound, as shown in the following equation.

$$2C + O_2 \longrightarrow 2CO$$

In addition, a synthesis reaction can involve the combining two compounds, as shown in the following equation.

$$CO_2 + H_2O \longrightarrow H_2CO_3$$

A second type of reaction is a **decomposition reaction**, in which a single compound breaks down to form one or more simpler substances.

$$CaCO_3 \longrightarrow CaO + CO_2$$

Self-Check What type of reaction is shown in the following equation?
$$2NH_3 \longrightarrow 3H_2 + N_2$$
Explain your answer.

A third type of reaction is called a **single replacement reaction**, where a single element reacts with a compound and displaces another element from the compound.

$$Zn + CuSO_4 \longrightarrow ZnSO_4 + Cu$$

Notice in the equation shown above that Zn replaces the Cu^{2+} ion in $CuSO_4$.

A fourth type of reaction is called a **double replacement reaction**, in which two atoms or ions appear to be exchanged between two compounds. Examine the following equation.

$$CdCl_2 + Na_2CO_3 \longrightarrow 2NaCl + CdCO_3$$

Notice that in the reaction shown above, the Cd^{2+} ion and the Na^+ ion appear to have exchanged places.

Unit VII Chemical Reactions continued

REVIEW YOUR UNDERSTANDING

_____ 10. Which chemical equation represents a single replacement reaction?
  (1) $Ca(ClO_3)_2 \rightarrow CaCl_2 + 3O_2$
  (2) $CaO + H_2O \rightarrow Ca(OH)_2$
  (3) $CaCO_3 + 2HCl \rightarrow CaCl_2 + H_2CO_3$
  (4) $Mg + 2AgNO_3 \rightarrow 2Ag + Mg(NO_3)_2$

_____ 11. Which type of reaction does the following chemical equation represent?

$$2HgO \rightarrow 2Hg + O_2$$

  (1) synthesis
  (2) decomposition
  (3) single replacement
  (4) double replacement

_____ 12. Which represents a correctly balance equation for a double replacement reaction?
  (1) $K_2SO_4 + Ba(NO_3)_2 \rightarrow KNO_3 + BaSO_4$
  (2) $Ca + Cl_2 \rightarrow CaCl_2$
  (3) $SO_2 + NO_2 \rightarrow SO_3 + NO$
  (4) $2KI + Pb(NO_3)_2 \rightarrow PbI_2 + 2KNO_3$

_____ 13. The reaction type that has only one reactant is called a
  (1) synthesis reaction.
  (2) decomposition reaction.
  (3) single replacement reaction.
  (4) double replacement reaction.

ANSWERS TO SELF-CHECK QUESTIONS

- The activated complex is located at the peak of the curve.

- The graph should look similar to the one for 2HI breaking apart to form H_2 and I_2. In this case, the energy level of the product, H_2O, must be lower than that of the reactants, H_2 and O_2.

- The three ways include raising the concentration of the reactants, increasing their surface area, and raising the temperature of the reaction mixture.

- A catalyst is a substance that speeds up the rate of a chemical reaction.

Unit VII Chemical Reactions continued

- $Al + CuCl_2 \longrightarrow AlCl_3 + Cu$

- $Al(s) + CuCl_2(aq) \longrightarrow AlCl_3(aq) + Cu(s)$

- There are 2 O atoms on the left side of the equation, but only 1 O atom on the right side. This imbalance makes it appear as if an O atom has been destroyed.

- $2Si + CO_2 \longrightarrow SiC + SiO_2$

- Both sides of the equation include 4 N atoms, 12 H atoms, and 10 O atoms.

- $3AgNO_3 + AlCl_3 \longrightarrow 3AgCl + Al(NO_3)_3$

- The equation represents a decomposition reaction because a compound is broken down into simpler substances.

Notes/Study Ideas/Answers

UNIT VII
Questions for Regents Practice

Holt Chemistry: The Physical Setting

PART A

_____ 1. What is conserved during a chemical reaction?
(1) energy, only
(2) mass, only
(3) both energy and mass
(4) neither energy nor mass

_____ 2. Which event must occur so that a reaction can happen?
(1) formation of a gas
(2) change of state
(3) addition of a catalyst to the reaction system
(4) effective collisions between reacting particles

_____ 3. Which two factors will increase the rate of a reaction?
(1) increasing the temperature and decreasing the concentration of the reactants
(2) decreasing the temperature and increasing the concentration of the reactants
(3) decreasing the temperature and decreasing the concentration of the reactants
(4) increasing the temperature and increasing the concentration of the reactants

_____ 4. In an exothermic reaction, the reactants have
(1) more energy than the products.
(2) the same amount of energy as the products.
(3) less energy than the products.
(4) the same amount of energy as supplied by the catalyst that is added to the reaction system.

_____ 5. In an equation, which symbol indicates that the substance is dissolved in water?
(1) (aq)
(2) (s)
(3) (l)
(4) (g)

_____ 6. If an equation is properly balanced, both sides of the equation must have the same number of
(1) coefficients.
(2) subscripts.
(3) atoms.
(4) molecules.

_____ 7. If all the oxygen atoms on one side of a balanced chemical equation are contained in the compound $Na_2CO_3 \cdot 10H_2O$, how many oxygen atoms must be present on the other side?
(1) 3
(2) 4
(3) 10
(4) 13

_____ 8. What type of reaction results in the formation of a compound that is more complex than the reactants?
(1) synthesis
(2) decomposition
(3) single replacement
(4) double replacement

Copyright © by Holt, Rinehart and Winston. All rights reserved.

Chemical Reactions

Unit VII Chemical Reactions continued

PART B-1

_____ 9. Examine the following reaction.

$$2C_4H_{10}(g) + 13O_2(g) \rightarrow 8CO_2(g) + 10H_2O(g) + energy$$

Which statement is true about the reaction shown above?
(1) The reactants and products include two states of matter.
(2) There are 20 H atoms on both sides of the equation.
(3) The products have more energy than the reactants.
(4) Both CO_2 and H_2O molecules must collide effectively for this reaction to occur.

_____ 10. Increasing the temperature increases the rate of a reaction by
(1) lowering the activation energy.
(2) increasing the activation energy.
(3) lowering the frequency of effective collisions between reacting molecules.
(4) increasing the frequency of effective collisions between reacting molecules.

_____ 11. The potential energy diagram below represents a chemical reaction.

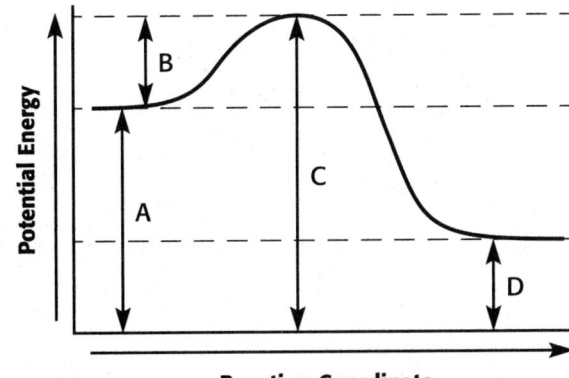

Which arrow represents the energy of the activated complex?
(1) A
(2) B
(3) C
(4) D

_____ 12. Examine the following reaction.

$$2Na + Cl_2 \rightarrow 2NaCl + energy$$

Which diagram best represents the potential energy changes for this reaction?

(1)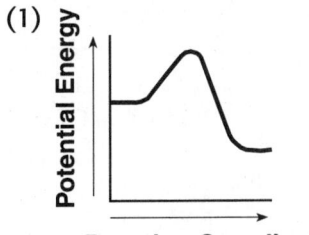

(2)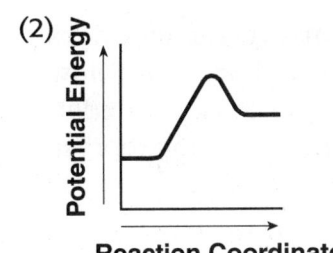

(3)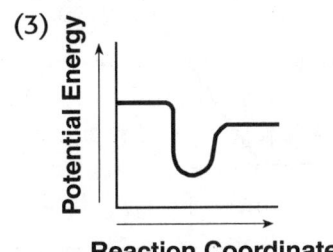

(4)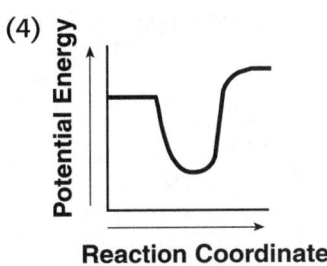

Unit VII Chemical Reactions continued

____ 13. Examine the following chemical equation.

$$H_2 + Fe_3O_4 \rightarrow Fe + H_2O$$

When the above equation is correctly balanced, the sum of the coefficients is
(1) 4.
(2) 8.
(3) 11.
(4) 12.

____ 14. Which equation is correctly balanced?
(1) $CaO + 2H_2O \rightarrow Ca(OH)_2$
(2) $2H_2 + O_2 \rightarrow 4H_2O$
(3) $2AlPO_4 + Ca(OH)_2 \rightarrow 2Al(OH)_3 + Ca_3(PO_4)_2$
(4) $Cl_2 + 2NaBr \rightarrow 2NaCl + Br_2$

PART B-2

Base your answers to questions 15-16 on the information and diagram below, which represent the changes in potential energy that occur during a reaction represented by the equation $A + B \rightarrow C$.

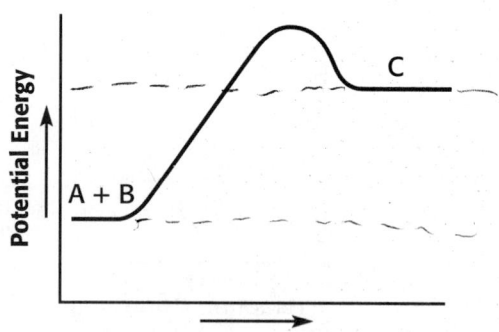

15. Does the diagram illustrate an exothermic or an endothermic reaction? Explain your answer.

16. Draw a dashed line on the graph to indicate a potential energy curve for the reaction when a catalyst is added.

17. Write the balanced equation for the double replacement reaction that occurs between ammonium chloride and silver nitrate.

Unit VII Chemical Reactions continued

PART C

18. Some antacid products contain Al(OH)₃ to neutralize HCl, which is the acid secreted by the stomach. These two compounds react to produce aluminum chloride and water. Write the balanced equation for this reaction. Explain why Al(OH)₃ would be a better choice than KOH.

19. The burning of wood is a chemical reaction. Wood is mostly cellulose, a compound that contains carbon, hydrogen, and oxygen. The ash that remains after the wood has been burned is mostly carbon and weighs much less than the original wood did. Explain why this observation does not violate the law of conservation of mass.

20. Methanol, CH₃OH, is a clean-burning fuel made by combining CO and H₂. Write the balanced equation for this synthesis reaction. Explain why methanol would be a clean burning fuel.

UNIT VIII
Stoichiometry

Holt Chemistry: The Physical Setting

A formula in a balanced equation indicates the chemical composition of each reactant and product. The coefficient before each formula indicates the proportion in which they react. Therefore, a balanced equation is similar to a recipe that tells you the proportion of each ingredient.

For example, consider the following synthesis reaction.

$$2H_2 + O_2 \longrightarrow 2H_2O$$

The coefficients in this equation indicate the relative proportions of each substance involved in this reaction. This equation can be read as "Two molecules of hydrogen gas react with one molecule of oxygen gas to form two molecules of water."

The coefficients from a balanced chemical equation can also be used to find the quantity of each reactant and product involved. The branch of chemistry that deals with the quantities of substances in chemical reactions is known as **stoichiometry**.

Moles

Quantity in chemistry is measured in a unit called the **mole**. A mole is defined as the number of atoms in exactly 12 grams of carbon-12. Chemists use the mole as a counting unit, just as you use the dozen as a counting unit. A chemist may require 2 moles of a substance, just as you may require 2 dozen cookies for a party.

Coefficients in a balanced equation also show the relative number of moles of the substances in the reaction. As a result, you can use the coefficients in conversion factors called **mole ratios**. Mole ratios allow you to convert from moles of one substance to moles of another substance. For example, consider the following chemical equation.

$$Fe_2O_3 + 2Al \longrightarrow 2Fe + Al_2O_3$$

The equation shown above can be read as "One mole of Fe_2O_3 reacts with two moles of Al to produce two moles of Fe and one mole of Al_2O_3." Therefore, the mole ratios in this equation are 1:2:2:1. This equation can also be read in terms of molecules: "One molecule of Fe_2O_3 reacts with two molecules of Al to produce two molecules of Fe and one molecule of Al_2O_3."

What You'll Need to Learn
This topic is part of the Regents Curriculum for the Physical Setting Exam.
Standard 4, Performance Indicator 3.3: *Apply the principle of conservation of mass to chemical reactions.*

What Terms You'll Need to Know
empirical formula
formula mass
gram-formula mass
mole
mole ratio
percent composition
stoichiometry

Where You Can Learn Even More:
Holt Chemistry: The Physical Setting
Chapter 3: Atoms and Moles (Section 4)
Chapter 7: The Mole and Chemical Composition
Chapter 9: Stoichiometry

Unit VIII Stoichiometry continued

Self-Check What are the mole ratios in the following equation?
$$2C_8H_{18} + 25O_2 \rightarrow 16CO_2 + 18H_2O$$

Mole ratios can be used to calculate the quantity of any substance in the reaction if you know the quantity of just one substance in the reaction. Consider the following reaction.

$$PCl_3 + 3H_2O \rightarrow H_3PO_4 + 3HCl$$

If you know how many moles of PCl_3 are used in this reaction, then you can calculate how many moles of H_2O are required and how many moles of H_3PO_4 and HCl are produced. For example, assume that you are asked to calculate how many moles of HCl will be produced if you has 24 moles of PCl_3. From the balanced equation, you know the mole ratio: 1 mol PCl_3: 3 mol HCl.

Use this mole ratio to solve the problem. Be sure to set up the mole ratio so that units cancel to leave the units required in the answer.

$$? \text{ mol HCl} = 24 \text{ mol } PCl_3 \times \frac{3 \text{ mol HCl}}{1 \text{ mol } PCl_3} = 72 \text{ mol HCl}$$

Self-Check How many moles of PCl_3 are required to react with 20 moles of H_2O?

In solving a stoichiometry problem using a mole ratio, be sure to the have the unknown substance on top and the substance given in the problem on the bottom so that the units cancel correctly.

REVIEW YOUR UNDERSTANDING
Given the following equation:

$$Al_2O_3 + 3C \rightarrow 2Al + 3CO$$

_____ 1. What is the mole ratio in this equation?
  (1) 1:3:3:2
  (2) 2:3:3:1
  (3) 1:3:2:3
  (4) 4:5

_____ 2. What is the mole ratio of CO to C?
  (1) 3:2
  (2) 1:1
  (3) 1:2
  (4) 2:3

Unit VIII Stoichiometry continued

____ 3. How many moles of Al will be produced if 6 moles of C react? Assume that Al_2O_3 is present in sufficient quantity.
(1) 2 mol Al
(2) 3 mol Al
(3) 4 mol Al
(4) 6 mol Al

____ 4. How many moles of Al_2O_3 are required so that all 6 moles of C do react?
(1) 1 mol C
(2) 2 mol C
(3) 3 mol C
(4) 6 mol C

Gram-Formula Mass

In chemistry, you often need to know the mass of a given number of moles of a substance. In some cases, you may need to know the number of moles given the mass of a substance. Calculating either the mass, given the number of moles, or the number of moles, given the mass, is straightforward because of the way the mole is defined.

What you must do in both types of problems is to calculate the formula mass of the substance. The **formula mass** is the sum of the atomic masses of all the atoms in a substance. You learned that the unit for atomic mass is the atomic mass unit (amu).

Determining the formula mass of an element is easy. All you have to do is check a periodic table. For example, the formula mass of C is 12.01 amu. Recall that some elements exist as substances that contain more than one atom. Examples include gaseous elements such as N_2, O_2, F_2, Cl_2, Br_2, and I_2. To calculate the formula mass of N_2, you must multiply the atomic mass of N by 2.

14.01 amu (formula mass of N) × 2 =
28.02 amu (formula mass of N_2)

Self-Check What is the formula mass of Cl_2?

Calculating the formula mass of a compound involves adding the formula masses of all the atoms present in the compound. For example, the formula mass of CO_2 is calculated as follows.

1 C atom = 12.01 amu
2 O atoms = 32.00 amu
CO_2 = 44.01 amu

Unit VIII Stoichiometry continued

Be sure to include the correct number of atoms when calculating the formula mass of a compound that contains a polyatomic ion, such as $(NH_4)_2SO_4$.

$$\begin{aligned} 2 \text{ N atoms} &= 28.02 \text{ amu} \\ 8 \text{ H atoms} &= 8.08 \text{ amu} \\ 1 \text{ S atom} &= 32.06 \text{ amu} \\ 4 \text{ O atoms} &= 64.00 \text{ amu} \\ \hline (NH_4)_2SO_4 &= 132.16 \text{ amu} \end{aligned}$$

Self-Check What is the formula mass of $Fe_2(CrO_4)_3$?

When dealing with quantities, chemists do not use the unit amu. Instead, they use the unit gram (g). The mass in grams of one mole of a substance is called the **gram-formula mass**, which is also called the *molar mass*. For example, you learned that the formula mass of one mole of $(NH_4)_2SO_4$ is 132.16 amu. Therefore, the gram-formula mass of $(NH_4)_2SO_4$ is 132.16 g/mol. Two moles of $(NH_4)_2SO_4$ would have a mass of 264.32 g.

Converting between moles and mass requires a conversion factor. The conversion factor used is the gram-formula mass. The following setup shows how to convert moles to mass.

$$\text{moles} \times \frac{\text{grams}}{1 \text{ mol}} = \text{grams}$$

The gram-formula mass of $Ba(NO_3)_2$ is 261.35 g/mol. The following shows how to calculate how many grams are present in 2.3 moles of $Ba(NO_3)$.

$$2.3 \text{ mol} \times \frac{26.35 \text{ g}}{1 \text{ mol}} = 601.11 \text{ g}$$

Self-Check Calculate the mass in grams of 0.5 moles of $Al_2(SO_4)_3$.

The following setup shows how to use the gram-formula mass to convert mass to moles.

$$\text{grams} \times \frac{1 \text{ mol}}{\text{grams}} = \text{moles}$$

You can then calculate how many moles are present in 259.35 g $Ca(OH)_2$. The gram-formula mass of $Ca(OH)_2$ is 74.10 g/mol.

$$259.35 \text{ g} \times \frac{1 \text{ mol}}{74.10 \text{ g}} = 3.5 \text{ mol}$$

Self-Check Calculate the number of moles present in 85.5 g of $Al_2(SO_4)_3$.

Unit VIII Stoichiometry continued

A gram-formula mass can also be used to calculate the **percent composition** of a substance. Because the gram-formula mass is expressed in grams, the percent composition indicates the percentages by mass.

For example, the percent of Cu by mass in Cu_2S can be calculated. You must determine what mass percent Cu contributes to the gram-formula mass of Cu_2S. To do this, you must first determine the gram-formula masses of Cu and Cu_2S.

A periodic table shows that Cu has an atomic mass of 63.55 amu and that S has an atomic mass of 32.07 amu. The gram-formula mass of Cu is 63.55 g/mol. The gram-formula mass of Cu_2S is 159.17 g/mol. Be sure to set up the problem to reflect that Cu_2S contains 2 Cu atoms.

$$\text{mass \% Cu} = \frac{2 \times 63.55 \text{ g/mol}}{159.17 \text{ g/mol}} \times 100 = 79.85\% \text{ Cu}$$

Remember to multiply a decimal value by 100 to change it to a percent.

Self-Check Use the gram-formula mass of Cu_2S to calculate the percent composition of S.

REVIEW YOUR UNDERSTANDING

Given the formula $Pb(ClO_3)_2$:

_____ **5.** What is the gram-formula mass of this compound?
  (1) 290.7 g/mol
  (2) 358.1 amu
  (3) 374.1 g/mol
  (4) 374.1 amu

_____ **6.** How many moles are present in 1234.5 grams of this compound?
  (1) 1 mol
  (2) 2.5 mol
  (3) 2.8 mol.
  (4) 3.3 mol

_____ **7.** How many grams are present in 0.6 mol of this compound?
  (1) 125 g
  (2) 175 g
  (3) 224 g
  (4) 600 g

Notes/Study Ideas/Answers

8. What is the percent composition of O in this compound?
(1) 19%
(2) 26%
(3) 55%
(4) 96%

Empirical and Molecular Formulas

You learned that a molecular formula indicates the number of atoms of each atom present in a compound. For example, the formula P_4O_{10} indicates that this compound consists of 4 P atoms and 10 O atoms. If you take another look at this formula, you will notice that the subscripts are not the simplest whole numbers.

The simplest ratio of the atoms in a compound is shown in an **empirical formula**. For example, the molecular formula P_4O_{10} can be rewritten as the empirical formula P_2O_5. Both kinds of formulas are just two different ways of representing the composition of the same molecule.

In general, the molecular formula is a whole-number multiple of the empirical formula. For example, the molecular formula P_4O_{10} is a multiple of 2 compared to the empirical formula P_2O_5.

> **Self-Check** Is $C_6H_{12}O_6$ an empirical formula or a molecular formula? Explain your answer.

If you know the empirical formula and the gram-formula mass of a compound, then you can determine its molecular formula. Consider a compound has the empirical formula CH and a gram-formula mass of 78 g/mol.

You must determine the factor that is used to multiply the empirical formula to arrive at the molecular formula. First, calculate the gram-formula mass of CH. For this type of problem, you can round the gram-formula masses of atoms to the nearest whole number.

$$\begin{aligned} 1 \text{ C atom} &= 12 \text{ g/mol} \\ 1 \text{ H atom} &= 1 \text{ g/mol} \\ \hline \text{CH} &= 13 \text{ g/mol} \end{aligned}$$

Next, divide the gram-formula mass of the compound by the gram-formula mass of the empirical formula. This will give you the multiple needed to get the molecular formula.

$$\frac{78 \text{ g/mol}}{13 \text{ g/mol}} = 6$$

Unit VIII Stoichiometry continued

Finally, multiply the empirical formula by this multiple to get the molecular formula: C_6H_6.

Self-Check What is the molecular formula of a compound that has the empirical formula NO_2 and a gram-formula mass of 92 g/mol?

Some compounds have the same molecular formula and empirical formula. Water, H_2O, is an example.

REVIEW YOUR UNDERSTANDING

_____ 9. Which is an empirical formula?
(1) H_2SO_4
(2) $C_4H_4N_2O_2$
(3) C_2H_4
(4) H_2O_2

_____ 10. The molecular formula for acetylene is C_2H_2. The molecular formula for benzene is C_6H_6. What is the empirical formula for both compounds?
(1) CH
(2) C_2H_2
(3) C_6H_6
(4) $(CH)_2$

_____ 11. You know the empirical formula of a compound. What other information do you need to calculate its molecular formula?
(1) the number of moles
(2) the number of grams
(3) the gram-formula mass
(4) the percent composition

_____ 12. The molecular formula of a compound is a multiple of 3 compared to its empirical formula, which is C_2H_2N. What is the molecular formula of this compound?
(1) $C_2H_2N_3$
(2) $C_4H_4N_2$
(3) C_6H_6N
(4) $C_6H_6N_3$

Unit VIII Stoichiometry continued

ANSWERS TO SELF-CHECK QUESTIONS

- The mole ratios are 2:25:16:18.
- 6.7 mol PCl_3
- 70.9 amu
- 459.7 amu
- 171.09 g
- 0.25 mol
- 20.15% S
- It is a molecular formula because it does not show the simplest whole-number ratio, which is CH_2O.
- N_2O_4

UNIT VIII
Questions for Regents Practice

Holt Chemistry: The Physical Setting

PART A

_____ 1. Stoichiometry problems often require the use of
   (1) Lewis structures.
   (2) energy diagrams.
   (3) a periodic table.
   (4) a chart of electron configurations.

_____ 2. Coefficients in chemical equations represent
   (1) gram-formula masses.
   (2) mole ratios.
   (3) empirical formulas.
   (4) the number of atoms present in each substance.

_____ 3. The number of atoms in exactly 12 grams of carbon-12 represents
   (1) 1 amu.
   (2) one molecule.
   (3) 1 mol.
   (4) one formula unit.

_____ 4. Given the following equation:

   $2NH_3 + O_2 \longrightarrow 2NO + 3H_2O$

   What is the mole ratio of O_2 to H_2O?
   (1) 1:3
   (2) 2:2
   (3) 2:3
   (4) 3:2

_____ 5. The formula mass of F_2 is
   (1) 19 amu.
   (2) 38 amu.
   (3) 19 g/mol.
   (4) 38 g/mol.

_____ 6. The gram-formula mass of F_2 is
   (1) 19 amu.
   (2) 38 amu.
   (3) 19 g/mol.
   (4) 38 g/mol.

_____ 7. The units for gram-formula mass are
   (1) g/mol.
   (2) mol/g.
   (3) atoms/mol.
   (4) atoms/g

_____ 8. To calculate the number of moles of a compound in a sample, you need to know its
   (1) empirical formula.
   (2) molecular formula.
   (3) mass and gram-formula mass.
   (4) percent composition.

_____ 9. The percent by mass of oxygen in CO_2 is equal to
   (1) $2/3 \times 100$
   (2) $2/32 \times 100$
   (3) $2/44 \times 100$
   (4) $32/44 \times 100$

Stoichiometry

Name _____ Class _____ Date _____

Unit VIII Stoichiometry continued

PART B-1

_____ 10. Examine the following reaction.

$$6CO_2(g) + 6H_2O(g) \rightarrow C_6H_{12}O_6(g) + 6O_2(g)$$

What is the total number of moles of water needed to make 4.5 moles of $C_6H_{12}O_6$?
(1) 4.5 mol
(2) 6 mol
(3) 18 mol
(4) 27 mol

_____ 11. What is the gram-formula mass of $CuSO_4$?
(1) 111.62 g/mol
(2) 138 g/mol
(3) 159.62 g/mol
(4) 206.5 g/mol

_____ 12. The gram-formula mass of NH_6Cl is
(1) 22.4 g/mol
(2) 28.0 g/mol
(3) 53.5 g/mol
(4) 95.5 g/mol

_____ 13. Which compound contains the greatest percent by mass of oxygen?
(1) BeO
(2) MgO
(3) CaO
(4) SrO

_____ 14. What is the molecular formula of a compound that has a molecular mass of 54 g/mol and the empirical formula C_2H_3?
(1) C_2H_3
(2) C_4H_6
(3) C_6H_9
(4) C_8H_{12}

_____ 15. A compound has the empirical formula CH_2. What is the gram-formula mass of a compound with a molecular formula that is three times its empirical formula?
(1) 14 g/mol
(2) 28 g/mol
(3) 36 g/mol
(4) 42 g/mol

PART B-2

16. Explain why it is necessary to use mole ratios in solving stoichiometry problems.

_____

_____

_____

17. Write the conversion factor needed to convert grams to moles.

_____

_____

_____

18. Calculate how many grams are present in 2.7 mol $NaNO_2$.

_____

_____

_____

Copyright © by Holt, Rinehart and Winston. All rights reserved.

Unit VIII Stoichiometry continued

PART C

19. TNT is an explosive whose formula is $C_6H_2CH_3(NO_2)_3$. Calculate the gram-formula mass of TNT.

20. Determine the percent composition of hydrogen and oxygen in TNT.

UNIT IX

Gases

Holt Chemistry: The Physical Setting

Compared to a solid and a liquid, a gas represents the state of matter where the particles have the greatest kinetic energy. As a result, the distance between the particles of a gas is much greater than the distance between the particles of a liquid or a solid. Because their particles are so far apart, gases have low density. They can also be compressed easily. Because of their high kinetic energy, gases completely fill a container, taking its shape and volume.

The Gas Laws

The behavior of a gas is a measurable property. As a result, its behavior can be described in terms of various laws. Each law is a mathematical description of how a gas behaves under various conditions, such as changes in pressure and temperature. As either of these conditions changes, so does the volume of a gas.

The relationship between pressure and volume of a gas is known as **Boyle's law**. Boyle's law states that this relationship is an inverse one. As the pressure on a gas increases, its volume decreases. Conversely, as the pressure on a gas decreases, its volume increases. Boyle's law operates only when the temperature and the number of particles of the gas are not changed.

Boyle's law is stated mathematically as follows.

$$P_1V_1 = P_2V_2$$

P_1V_1 represents the pressure and volume of the gas under one set of conditions, while P_2V_2 represents the pressure and volume of the gas under another set of conditions. For example, assume a given sample of gas occupies 436 mL at 760 mm of pressure. Using Boyle's law, you can calculate what the volume of this gas will be if the pressure is increased to 820 mm. Both the number of particles of the gas and temperature are kept constant.

$$P_1 = 760 \text{ mm} \quad V_1 = 436 \text{ ml} \quad P_2 = 820 \text{ mm} \quad V_2 = ?$$
$$P_1V_1 = P_2V_2$$
$$(760 \text{ mm})(436 \text{ mL}) = (820 \text{ mm})V_2$$
$$V_2 = \frac{(760 \text{ mm})(436 \text{ mL})}{820 \text{ mm}} = 404 \text{ mL}$$

What You'll Need to Learn

This topic is part of the Regents Curriculum for the Physical Setting Exam.
Standard 4, Performance Indicator 3.4: *Use kinetic molecular theory (KMT) to explain rates of reactions and the relationships among temperature, pressure, and volume of a substance.*

What Terms You'll Need to Know

Boyle's law
Charles's law
ideal gas
kinetic molecular theory (KMT)
standard temperature and pressure

Where You Can Learn Even More:

Holt Chemistry: The Physical Setting
Chapter 12: Gases

Self-Check A sample of gas has a volume of 160 mL at a pressure of 760 mm. What must be the pressure if the same sample of gas occupies 210 mL at the same temperature?

Unit IX Gases continued

The behavior of a gas can also be studied when the temperature is changed while the pressure is kept constant. The relationship between temperature and volume of a gas is known as **Charles's law**. Charles's law states that this relationship is a direct one. As the temperature on a gas increases, its volume increases. Conversely, as the temperature on a gas increases, its volume increases. Charles's law operates only when the pressure and the number of particles of the gas are not changed.

Charles's law is stated mathematically as follows.

$$\frac{V_1}{T_1} = \frac{V_2}{T_2}$$

When using Charles's law, the unit used for temperature must be Kelvins (K). If the temperature is given in degrees Celsius, then the following formula is used to convert to K.

$$K = °C + 273$$

Therefore, 28°C is equal to 301 K. To change from K to °C, you would subtract 273 K. These changes in temperature values must be performed when using Charles's law. For example, assume that a balloon is inflated to 730 mL volume at 28°C. Suppose the balloon is immersed in a dry-ice bath at −80°C. What will be the volume of the balloon, assuming that the number of particles of gas and the pressure are kept constant?

$$V_1 = 730 \text{ mL} \quad T_1 = 301 K \quad V_2 = ? \quad T_2 = 193 \text{ K}$$

$$\frac{V_1}{T_1} = \frac{V_2}{T_2}$$

$$\frac{730 \text{ mL}}{301 \text{ K}} = \frac{V_2}{193 \text{ K}}$$

$$V_2 = \frac{730 \text{ mL}(193 \text{ K})}{301 \text{ K}} = 468 \text{ mL}$$

> **Self-Check** A sample of gas has a volume of 368 mL at a temperature of 22°C. What must be the temperature if the same sample of gas occupies 539 mL at the same pressure?

The effects of changing the pressure and temperature of a sample of gas are sometimes compared to a set of standard conditions. These conditions are known as **standard temperature and pressure** (STP), which is equal to 0°C and 760 mm. *Reference Table A* lists the values and units for STP. This table includes another unit used for pressure—kilopascal (kPa). Problems in this book will use only *mm* as the unit for pressure.

Unit IX Gases continued

In 1811, an Italian scientist named Amadeo Avogadro proposed the idea that equal volumes of all gases that are under the same conditions of temperature and pressure have the same number of particles. This proposal was later shown to be correct and became known as Avogadro's law.

REVIEW YOUR UNDERSTANDING

_____ 1. Boyle's law deals with the relationship between
   (1) volume and temperature.
   (2) volume and pressure.
   (3) temperature and pressure.
   (4) volume and the number of particles.

_____ 2. Charles's law involves the relationship between
   (1) volume and temperature.
   (2) volume and pressure.
   (3) temperature and pressure.
   (4) volume and the number of particles.

_____ 3. According to Boyle's law, as the pressure on a gas increases, its
   (1) volume decreases.
   (2) volume increases.
   (3) temperature decreases.
   (4) temperature increases.

_____ 4. Which represents 25°C in Kelvins (K)?
   (1) 25 K
   (2) 248 K
   (3) 273 K
   (4) 298 K

_____ 5. Charles's law shows a direct relationship because the
   (1) volume of a gas increases as the pressure is increased.
   (2) volume of a gas decreases as the pressure is increased.
   (3) volume of a gas is increased as the temperature is increased.
   (4) the number of particles is the same for all gases that exist under the same conditions.

Kinetic Molecular Theory

The behavior of gases is explained at the molecular level in terms of the **kinetic-molecular theory**. This theory includes four statements about the behavior of gases.

1. *Gas particles are in constant, random motion.* This explains why gases have no fixed shape or volume. You learned that the particles in a solid remain in fixed positions, while the particles in a liquid can move, although not as freely as those in a gas.

2. *The particles of a gas are very far apart relative to their size.* This relatively large distance between particles explains why gases can be easily compressed.

3. *Gas particles do not have any attractive force for one another.* This lack of attraction explains why gases take the shape of any container they occupy.

4. *Collisions between gas particles are perfectly elastic.* This means that gas particles do not lose any energy when they collide. Instead, the energy is completely transferred between particles during a collision. As a result, the total energy of the system remains constant.

Self-Check Explain how the kinetic-molecular theory is appropriately named.

You examined some of the laws that apply to gases. These laws can be explained in terms of the kinetic molecular theory. However, no gas obeys all these laws under all conditions. As a result, the best way to develop a model of gas behavior is to assume that the gas behaves like an ideal gas. An ideal gas is one that behaves exactly the way the kinetic-molecular theory predicts that it should.

For example, the kinetic-molecular theory states that gas particles have no forces of attraction for one another. However, gas particles do attract, and even repulse, one another. These forces are very weak, especially when compared to those that exist in solids and liquids.

Gases act most like an ideal gas when two conditions are met. These two conditions include low pressure and high temperature. These two conditions cause the gas particles to be farthest apart from one another. As a result, any attractive or repulsive force between gas particles is minimized.

Name _____ Class _____ Date _____

Unit IX Gases *continued*

REVIEW YOUR UNDERSTANDING

Notes/Study Ideas/Answers

_____ 6. The kinetic-molecular theory states that ideal gas molecules
   (1) have weight and take up space.
   (2) are in constant, rapid, and random motion.
   (3) exert forces of attraction and repulsion on one another.
   (4) have high densities compared to liquids and solids.

_____ 7. Which two conditions favor a gas behaving like an ideal gas?
   (1) low temperature and low pressure
   (2) low temperature and high pressure
   (3) high temperature and low pressure.
   (4) high temperature and high pressure

_____ 8. In an ideal gas, two particles that collide
   (1) lose energy.
   (2) gain energy.
   (3) transfer energy.
   (4) slow down.

_____ 9. Gas particles completely fill any container despite its shape because they
   (1) move at the same speed at all temperatures.
   (2) resist changes in pressure.
   (3) occupy the same volume at all temperatures and pressures.
   (4) have weak forces of attraction for one another.

ANSWERS TO SELF-CHECK QUESTIONS

- 579 mm

- 432 K = 159°C

- The term *kinetic* refers to motion, as illustrated in the concept of kinetic energy. The term *molecular* refers to the gas particles (molecules) whose behavior the theory explains.

UNIT IX
Questions for Regents Practice

Holt Chemistry: The Physical Setting

PART A

_____ 1. According to Boyle's law,
(1) increasing the pressure increases the volume of a gas.
(2) decreasing the pressure increases the volume of a gas.
(3) increasing the temperature increases the volume of a gas.
(4) decreasing the temperature increases the volume of a gas.

_____ 2. Which sample of NH_3 will completely fill and take the shape of a closed container?
(1) $NH_3(s)$
(2) $NH_3(l)$
(3) $NH_3(g)$
(4) $NH_3(aq)$

_____ 3. Which graph shows the relationship between temperature and pressure for an ideal gas?

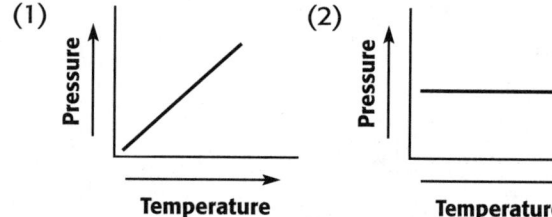
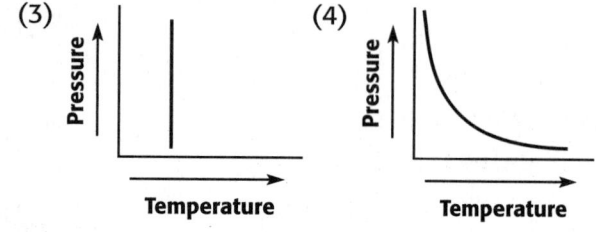

_____ 4. At the same temperature and pressure, which sample contains the same number of moles of particles as 1 L of CO_2?
(1) 2 L $CO_2(g)$
(2) 0.5 L $O_2(g)$
(3) 1 L $H_2O(l)$
(4) 1 L $Ar(g)$

_____ 5. A real gas behaves more like an ideal gas when the gas molecules are
(1) far apart and have strong attractive forces between them.
(2) are far apart and have weak attractive forces between them.
(3) close and have strong attractive forces between them.
(4) close and have weak attractive forces between them.

_____ 6. Unlike an ideal gas, a real gas has molecules that
(1) take up a larger volume when the pressure is increased.
(2) have very low kinetic energy.
(3) have forces of attraction for one another.
(4) do not collide with one another.

_____ 7. Standard temperature and pressure are defined as
(1) the conditions under which a real gas behaves most like an ideal gas.
(2) the ideal conditions for a gas to occupy the largest volume.
(3) 0 K and 760 mm, respectively.
(4) 0°C and 760 mm, respectively.

Unit IX Gases continued

PART B-1

___ 8. A gas has a volume of 275 mL at a pressure of 760 mm and a temperature of 25°C. Which will increase the volume of this gas?
(1) changing the pressure to 900 mm
(2) changing the pressure to 700 mm
(3) changing the temperature to 15°C
(4) changing the temperature to 273 K

___ 9. A gas occupies of volume of 50 mL at 35°C. If the volume is increased to 75 mL at a constant pressure, the resulting temperature will be equal to
(1) $35°C \times \dfrac{50 \text{ mL}}{75 \text{ mL}}$
(2) $35°C \times \dfrac{75 \text{ mL}}{50 \text{ mL}}$
(3) $308 \text{ K} \times \dfrac{50 \text{ mL}}{75 \text{ mL}}$
(4) $308 \text{ K} \times \dfrac{75 \text{ mL}}{50 \text{ mL}}$

___ 10. A gas occupies of volume of 150 mL at 780 mm. If the volume is decreased to 110 mL at a constant temperature, the resulting pressure will be equal to
(1) $780 \text{ mm} \times \dfrac{150 \text{ mL}}{110 \text{ mL}}$
(2) $780 \text{ mm} \times \dfrac{110 \text{ mL}}{150 \text{ mL}}$
(3) $110 \text{ mL} \times \dfrac{150 \text{ mL}}{780 \text{ mm}}$
(4) $780 \text{ mm} \times 150 \text{ mL} \times 110 \text{ mL}$

___ 11. Under which conditions will a sample of gas occupy the largest volume?
(1) standard temperature and pressure
(2) 273 K and 760 mm
(3) 250 K and 1000 mm
(4) 300 K and 300 mm

___ 12. Based on Charles's law, doubling the temperature of a sample of gas kept at a constant pressure will
(1) decrease its volume in half.
(2) double its volume.
(3) not affect the volume of the gas as long as the pressure remains constant.
(4) have the same effect as doubling the pressure on the gas at a constant temperature.

___ 13. Which two substances have the same number of particles at STP?
(1) 1 L of $H_2O(l)$ and 1 L $H_2O(g)$
(2) 1 L $O_2(g)$ and 0.5 L $CO_2(g)$
(3) 2 L $NH_3(g)$ and 2 L $CH_4(g)$
(4) 1 L $H_2O(g)$ and 1 kg $H_2O(s)$

___ 14. The smaller the gram-formula mass of a gas, the more likely it is to behave like an ideal gas. Therefore, which two gases are most likely to behave like an ideal gas?
(1) H_2 and He
(2) O_2 and H_2
(3) N_2 and O_2
(4) He and Ne

Unit IX Gases continued

_____ 15. The smaller the gram-formula mass of a gas, the more quickly it will diffuse. Which gas will diffuse the fastest?
(1) O_2
(2) CO_2
(3) CH_4
(4) Xe

PART B-2

16. Draw a graph to show the relationship between the volume of an ideal gas and its pressure at constant temperature.

17. Draw a graph to show the relationship between the volume of an ideal gas and its temperature at constant pressure.

18. Give two reasons why real gases do not behave exactly according to Boyle's law and Charles's law.

_____

_____

_____

PART C

19. A diver underwater releases a gas bubble with a volume of 100 mL at a depth where the pressure is 8360 mm. What will be the volume of this gas bubble when it reaches the surface where the pressure is 760 mm?

20. A person breathes 2.6 L of cold air into her lungs. The air is warmed from its initial temperature of –11°C to 37°C. What is the volume of this air after it is warmed in her lungs?

UNIT X
Solutions

Holt Chemistry: The Physical Setting

You learned that elements and compounds are pure substances that have a definite composition regardless of the size and location in a sample. In contrast, mixtures vary in composition. A homogeneous mixture is also known as a **solution**. For example, dissolving sugar in water makes a solution. The sugar molecules are evenly dispersed throughout the water to form a homogeneous mixture.

Solvents, Solutes, and Solubility

A solution is composed of particles of two or more substances evenly distributed amongst each other. The main ingredient in a solution is called a **solvent**. Water is the most common solvent. Although it is very common, water is unique because so many substances can dissolve in it. Solutions with water as the solvent are called aqueous solutions. You learned that the symbol *aq* is used to show a substance that exists in an aqueous solution.

A substance that dissolves in a solvent is called a **solute**. Sugar is a solute that dissolves easily in water. However, sand is not a solute because it does not dissolve in water no matter how long you stir it.

Some substances, such as sugar, dissolve easily. Other substances do not dissolve as easily or quickly. Some substances, such as sand, do not dissolve at all. The extent to which a substance dissolves in a solvent such as water is referred to as its **solubility**.

A highly soluble substance dissolves easily in a solvent. A slightly soluble substance dissolves only to a limited extent in a solvent. An insoluble substance does not dissolve to any measurable extent.

Knowing the chemical composition of the substances may help you predict if they will form a solution, and if so, how well the solute will dissolve in the solvent. The general rule is "like dissolves like." This rule means that a polar substance generally dissolves in another polar substance, while a nonpolar substance generally dissolves in another nonpolar substance.

Most solutions are made by dissolving a solid in a liquid. Many ionic compounds are highly soluble in water. One example is NaCl. You learned that NaCl consists of Na^+ and Cl^- ions. As a polar molecule, water can break apart, or dissociate, a Na^+ ion from a Cl^- ion.

What You'll Need to Learn
This topic is part of the Regents Curriculum for the Physical Setting Exam.
Standard 4, Performance Indicator 3.1: *Explain the properties of materials in terms of the arrangement and properties of the atoms that compose them.*

What Terms You'll Need to Know
concentration
electrolyte
molarity
saturated solution
solubility
solute
solution
solvent
supersaturated solution
unsaturated solution

Where You Can Learn Even More:
Holt Chemistry: The Physical Setting
Chapter 13: Solutions

Unit X Solutions continued

Self-Check What is the difference between a solute and a solvent?

The following models illustrate how NaCl, the solute, dissociates in water, the solvent, to form a solution.

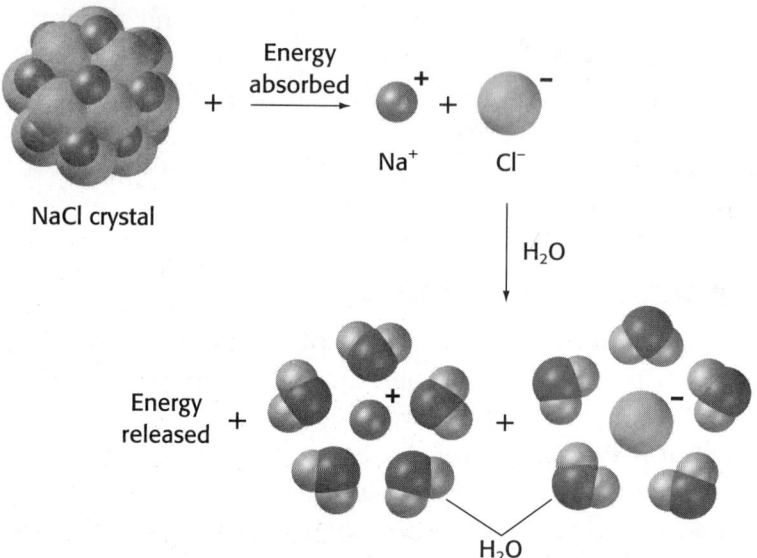

Notice that the ions must first be dissociated from their bonded positions in the salt crystal. This step requires energy. This energy is provided by the polar ends of water molecules that approach the ions and release energy. This attractive force is known as a molecule-ion attraction.

Notice that the dissociated ions are then hydrated, which means that water molecules surround them. The hydration step releases energy

The energy that is required to dissociate the ions and the energy that is released when the ions become hydrated depend on the cations and anions that make up the salt. The net energy gain or loss determines the solubility of an ionic solid.

The best way to determine the solubility of a substance is to measure it experimentally. These values can then be organized into a table, such as *Reference Table F*. This table shows which ionic compounds will dissolve in water to form an aqueous solution. This table also lists exceptions to the general rule.

For example, *Reference Table F* indicates that compounds that contain the polyatomic ion, NO_3^-, are soluble. Those that contain the polyatomic ion, SO_4^{2-} are also soluble unless this ion is combined with Ag^+, Ca^{2+}, Sr^{2+}, Ba^{2+}, and Pb^{2+}.

Self-Check Check *Reference Table F* to determine if the following ionic compounds are soluble: $AgHCO_3$, NH_4ClO_4, $CaCl_2$, and PbI_2.

Unit X Solutions *continued*

Notice that *Reference Table F* also lists ionic compounds that are insoluble. For example, compounds containing the polyatomic ion CO_3^{2-} are insoluble unless this ion is combined with a Group 1 metal or NH_4^+. Therefore, $CaCO_3$ is insoluble, while Na_2CO_3 is soluble.

Self-Check Check *Reference Table F* to determine if the following ionic compounds are soluble: $AgCO_3$, NH_4PO_4, CaS_2, and $NaOH$.

More detailed information about the solubility of a compound is provided with the use of a solubility curve. A solubility curve indicates to what extent a solute dissolves in a solvent at a given temperature. For example, *Reference Table G* plots the number of grams of solute dissolved per 100 g of H_2O versus increasing temperature (°C). Notice that *Reference Table G* table plots the solubility data for ten different compounds.

If you examine *Reference Table G*, you will observe that most of the compounds display a greater solubility as the temperature is increased. For example, 80 g of $NaNO_3$ dissolve in 100 g of H_2O at 10°C. However, 140 g of $NaNO_3$ dissolve in this same amount of water at about 74°C.

Self-Check Compare the solubility of KNO_3 at 10°C and at 70°C.

Generally, the solubility of solids in water increases with higher temperatures. However, the solubility of some solids does not change significantly with an increase in temperature. For example, if you again check *Reference Table G*, you will see that the solubility of NaCl increases only about 2 g per 100 g of H_2O even though the temperature is increased from 0°C to 100°C.

The solubility of some compounds even decreases with an increase in temperature. Three such compounds are shown on *Reference Table G*: HCl, NH_3, and SO_2. Both NH_3 and SO_2 are gases. The solubility of gases generally decreases as the temperature increases.

Consider what happens if you were to open a bottle of warm carbonated soda and a bottle of cold carbonated soda at the same time. The warm soda will taste flat more quickly than the cold soda. Warm soda tastes flat because there is less CO_2 dissolved in it. Gases are less soluble in a liquid of higher temperature. The higher the temperature is, the higher the average kinetic energy of the gas molecules. As a result, the gas molecules are more likely to escape from the solvent.

Unit X Solutions continued

Self-Check Is the solubility of NH_3 or SO_2 affected more by increasing the temperature from 10°C to 40°C?

The solubility of gases in a liquid is also affected by changes in pressure. Generally, increasing the pressure increases the solubility of a gas. Think about what happens when you open a bottle of soda. Removing the cap reduces the pressure on the liquid. The reduced pressure allows the CO_2 molecules to escape from the liquid, causing the soda to taste flat.

REVIEW YOUR UNDERSTANDING

_____ 1. A solution is a
   (1) pure substance.
   (2) homogeneous mixture.
   (3) heterogeneous mixture.
   (4) mixture of an insoluble salt and water.

_____ 2. Water is considered a universal solvent because
   (1) only a few substances dissolve in it.
   (2) so many substances dissolve in it.
   (3) all ionic compounds dissolve in it.
   (4) water exists everywhere on Earth.

_____ 3. As the temperature increases, the solubility of a solid in water generally
   (1) remains the same.
   (2) decreases.
   (3) increases.
   (4) doubles for every 10°C rise in temperature.

_____ 4. Decreasing the pressure on a solution that contains a gas as a solute
   (1) decreases the solubility of a gas.
   (2) increases the solubility of a gas.
   (3) has the same effect as increasing the temperature.
   (4) has no effect on the solubility of the gas.

_____ 5. Which compound is most soluble at 60°C?
   (1) NH_4Cl
   (2) SO_2
   (3) NaCl
   (4) KNO_3

Unit X Solutions continued

Properties of Solutions

You learned that a solubility curve indicates the amount of a solute that dissolves in a given amount of solvent at a particular temperature. For example, according to *Reference Table G*, 50 g of NH_4Cl dissolve in 100 g of H_2O at 45°C. This is the maximum amount of NH_4Cl that will dissolve in this amount of water at this temperature. When the maximum amount of a solute is dissolved in a solution, the solution is said to be a **saturated solution**.

If a solution does not contain the maximum amount of the solute, then it is an **unsaturated solution**. In this case, more solute can be dissolved in the solution.

Under certain conditions, more solute can be dissolved than is indicated by a solubility curve. A solution that contains more solute dissolved than the solubility indicates is a **supersaturated solution**.

> **Self-Check** How many grams of KNO_3 must be dissolved in 100 g of H_2O at 50°C for the solution to be considered supersaturated?

Solutions are often described in terms of their **concentration**. The concentration indicates how much solute is present in a given volume of solution. One way to describe the concentration of a solution is to use the term unsaturated, saturated, or supersaturated. For example, if you were told that a solution is saturated, then you know something about its concentration. You know that the solution contains the maximum amount of solute that can dissolve at that temperature.

The concentration of a solution can be expressed in other ways. For example, the concentration may be given as parts of solute per million parts of solvent (ppm).

The concentration of a solution can also be described in terms of its **molarity**. Molarity (M) is a concentration unit expressed as moles of solute dissolved per liter of solution. Molarity describes how many moles of solute are in a liter of solution.

For example, a 1M NaCl solution contains 1 mol NaCl dissolved in 1 L of water, while a 2M NaCl solution contains 2 mol NaCl dissolved in 1 L of water. The definition of molarity can be stated as the following equation.

$$\text{molarity} = \frac{\text{moles of solute}}{\text{liter of solution}}$$

You can use this equation to calculate any one value if you are given the other two values. For example, assume that you

Unit X Solutions continued

are asked to calculate how many moles of solute are contained in 400 mL of a 2 M solution.

$$2\text{ M} = \frac{?\text{ moles}}{0.4\text{ L}}$$

$$?\text{ moles} = (2\text{ M})(0.4\text{ L}) = 0.8\text{ moles}$$

Notice that you must use the liter unit (L) for volume when working with the equation for molarity. Therefore, 400 mL must be converted to 0.4 L.

Self-Check In what volume of solvent must 4.5 moles of solute be dissolved to make a 9 M solution?

Dissolving a solute not only affects the concentration of a solution but can also affect other physical properties. For example, some solutions can conduct electricity, while others cannot. The ability to conduct electricity depends on the chemical nature of the solute that is dissolved in the solution.

Some solutes are **electrolytes**. An electrolyte is a substance that dissolves in water to give a solution that conducts an electric current. A solution can conduct electricity if it contains charged particles that are free to move through the solvent. Soluble ionic compounds make good electrolytes.

For example, dissolving NaCl(s) in water produces a solution that is an excellent conductor of electricity.

$$NaCl(s) \longrightarrow Na^+(aq) + Cl^-(aq)$$

Notice that the ionic solid dissociates so that the Na^+ and Cl^- ions can move about in the water.

Unlike sodium chloride, sucrose, or table sugar, is not an electrolyte. Sucrose, $C_{12}H_{22}O_{11}$, is a molecular compound that dissolves in water. However, sucrose does not dissociate in water to produce ions. Therefore, a solution containing dissolved sucrose has no charged particles to conduct electricity.

Self-Check Why is sucrose considered a nonelectrolyte?

Solutes can also affect the freezing and boiling points of a solution. Solutes generally lower the freezing point. This is the reason why salt is tossed on icy sidewalks. The salt actually lowers the freezing point of water. Therefore, the ice melts at a lower temperature than it normally would.

Solutes have the opposite effect on the boiling point of a solution. Solutes raise the boiling point. This is the reason why glycol is added to a car's radiator. The glycol lowers the freez-

Unit X Solutions continued

ing point of water so that it will not freeze in winter. The glycol also raises the boiling point of water so that the radiator will not overheat in summer. The more glycol that is added, the more the freezing point is lowered, and the more the boiling point is raised.

REVIEW YOUR UNDERSTANDING

_____ 6. A solution that contains the maximum amount of solute at a given temperature is said to be
   (1) unsaturated.
   (2) saturated.
   (3) supersaturated.
   (4) concentrated.

_____ 7. Molarity is expressed in units of
   (1) moles of solute per liter of solvent.
   (2) moles of solute per liter of solution.
   (3) liters of solvent per mole of solvent.
   (4) liters of solution per mole of solvent.

_____ 8. The addition of a solute, such as NaCl,
   (1) lowers the boiling point and lowers the freezing point of the solution.
   (2) lowers the boiling point and raises the freezing point of the solution.
   (3) raises the boiling point and lowers the freezing point of the solution.
   (4) raises the boiling point and raises the freezing point of the solution.

_____ 9. A solution can conduct an electric current if it contains a(n)
   (1) solute.
   (2) solvent.
   (3) molecular compound.
   (4) electrolyte.

_____ 10. Concentration refers to the
   (1) amount of solute present in the solvent.
   (2) maximum amount of solute that can be dissolved in a solvent.
   (3) amount of solute added to lower the freezing point of the solution by 10°C.
   (4) the amount of electrolytes added to the solvent.

Unit X Solutions continued

ANSWERS TO SELF-CHECK QUESTIONS

- A solute is the substance that is dissolved in the solvent. The solvent is the substance in which the solute is dissolved.

- $AgHCO_3$ (soluble); NH_4ClO_4 (soluble); $CaCl_2$ (soluble); PbI_2 (insoluble)

- $AgCO_3$ (insoluble); NH_4PO_4 (soluble); CaS_2 (insoluble); $NaOH$ (soluble)

- At 10°C, about 24 g of KNO_3 can dissolve in 100 g of H_2O, while about 135 g of KNO_3 can dissolve in the same amount of water at 70°C.

- The solubility of NH_3 decreases more significantly than the solubility of SO_2 as the temperature is raised.

- More than 85 g of KNO_3 must be dissolved in 100 g of H_2O for the solution to be considered supersaturated at 50°C.

- 500 mL or 0.5 L

- A solution containing dissolved sucrose does not conduct electricity.

Name _____ Class _____ Date _____

Holt Chemistry: The Physical Setting

UNIT X

Questions for Regents Practice

PART A

_____ 1. Which of the following *cannot* be considered a solution?
(1) soda
(2) distilled (pure) water
(3) tap water
(4) salt water

_____ 2. Which of the following *cannot* be a solvent?
(1) alcohol
(2) milk
(3) $H_2O(l)$
(4) $H_2O(s)$

_____ 3. Water is best described as
(1) a universal solute.
(2) a solution.
(3) a universal solvent.
(4) an electrolyte.

_____ 4. A substance that conducts an electric current when dissolved in water is called a(n)
(1) solvent.
(2) electrolyte.
(3) salt.
(4) ionic compound.

_____ 5. When NaCl is added to water, the solution will have a
(1) higher freezing point and a lower boiling point than water.
(2) higher freezing point and a higher point than water.
(3) lower freezing point and a higher boiling point than water.
(4) lower freezing point and a lower boiling point than water.

_____ 6. A 1 M solution contains
(1) one mole of solvent in one mole of solvent.
(2) one mole of solute in 1 L of solvent.
(3) one mole of solute in 1 L of solution.
(4) more solute than is normally present at a given temperature and pressure.

_____ 7. As the temperature decreases, the solubility of a solute in a solvent generally
(1) increases.
(2) remains the same.
(3) decreases.
(4) approaches its maximum value.

_____ 8. To calculate the molarity of a solution, you need to know the
(1) number of moles of solute and volume of solution.
(2) number of moles of solute and temperature.
(3) volume of solvent and temperature.
(4) chemical composition of the solute.

Copyright © by Holt, Rinehart and Winston. All rights reserved.

Solutions   Focus On Regents Exam

Name _____ Class _____ Date _____

Unit X Solutions continued

PART B-1

_____ 9. Which ionic compound is *soluble* in water?
  (1) AgCl
  (2) AgBr
  (3) Ag_2SO_4
  (4) $AgC_2H_3O_2$

_____ 10. Which ionic compound is *insoluble* in water?
  (1) K_2CO_3
  (2) K_2CrO_4
  (3) $(NH_4)_3PO_4$
  (4) $Ca_3(PO4)_2$

_____ 11. What is the maximum number of grams of $NaNO_3$ that can dissolve in 100 g of H_2O at 10°C?
  (1) 10 g
  (2) 40 g
  (3) 60 g
  (4) 80 g

_____ 12. At 90°C, 10 g of a substance saturates 100 g of water. The substance could be
  (1) NH_4Cl.
  (2) HCl.
  (3) $KClO_3$.
  (4) NH_3.

_____ 13. Which compound is most soluble in water at 50°C?
  (1) KCl
  (2) KNO_3
  (3) HCl
  (4) $KClO_3$

_____ 14. At what temperature is 100 g of water saturated when it contains 60 g NH_4Cl?
  (1) 20°C
  (2) 35°C
  (3) 50°C
  (4) 65°C

_____ 15. Which contains the greatest number of moles of solute?
  (1) 2 L of 0.5 M solution
  (2) 0.5 L of 2 M solution
  (3) 0.5 L of 0.5 M solution
  (4) 2 L of 2.0 M solution

_____ 16. What is the molarity of a solution than contains 1.5 mol $CaCl_2$ in 0.5 L of solution?
  (1) 1.5 M
  (2) 2.0 M
  (3) 3.0 M
  (4) 3.5 M

PART B-2

17. A solution contains 0.75 mol of solute dissolved in 750 mL of solvent. What is the molarity of this solution?

18. A solution is made by dissolving 5.6 g NaCl in 50.0 mL of water. Calculate the molarity of this solution. Hint: First, change mass to moles using the gram-formula mass of NaCl.

Unit X Solutions continued

PART C

19. Many reagent chemicals used in a lab are concentrated aqueous solutions. For example, sulfuric acid is usually 18 M H_2SO_4. How many moles of this compound are present in 25 mL of solution?

20. Draw a graph to show what happens over time to the solubility of CO_2 in a carbonated soda after the cap is removed.

Name _____ Class _____ Date _____

Holt Chemistry: The Physical Setting

UNIT XI
Acids and Bases

You learned that some solutes are electrolytes because they produce solutions that conduct electricity. One group of electrolytes includes acids. Another group includes bases. As electrolytes, both acids and bases share one property. They both dissociate in water to form ions that can conduct electricity. Those that dissociate completely into ions are strong electrolytes. In contrast, those that dissociate only partially to produce relatively few ions are considered weak electrolytes.

Identifying Acids and Bases

The definitions of an acid and a base have changed as chemists learned more about these substances. In 1890, Svante Arrhenius, a Swedish chemist, defined an acid as any substance that, when added to water, produces a hydrogen ion (H^+). Today, scientists recognize that an acid produces a **hydronium ion,** H_3O^+, in aqueous solutions. A hydronium ion is produced when a hydrogen ion combines with a water molecule as shown in the following equation.

$$H^+(aq) + H_2O(l) \longrightarrow H_3O^+(aq)$$

Therefore, an **Arrhenius acid** is any substance that produces H^+ or H_3O^+ ions when added to water. A strong acid is one that dissociates in water to produce a high concentration of H^+ or H_3O^+ ions. For example, nitric acid, HNO_3, reacts with water as shown in the following equation.

$$HNO_3(l) + H_2O(l) \longrightarrow H_3O^+(aq) + NO_3^-(aq)$$

Nitric acid completely dissociates so that no HNO_3 molecules are present in solution. In contrast, acetic acid, CH_3COOH, is a weak acid because it only partially dissociates to produce relatively few H^+ or H_3O^+ ions. The following equation shows the reaction of acetic acid with water.

$$CH_3COOH(l) + H_2O(l) \longrightarrow H_3O^+(aq) + CH_3COO^-(aq)$$

In the above reaction, acetic acid still remains in solution.

What You'll Need to Learn

This topic is part of the Regents Curriculum for the Physical Setting Exam.
Standard 4, Performance Indicator 3.1: *Explain the properties of materials in terms of the arrangement and properties of the atoms that compose them.*

Which Terms You'll Need to Know

acid
Arrhenius acid
Arrhenius base
base
hydronium ion
indicator
neutralization reaction
pH
titration

Where You Can Learn Even More

Holt Chemistry: The Physical Setting
Chapter 15: Acids and Bases

Unit XI Acids and Bases continued

One molecule of some acids can react to form more than one hydronium ion. An example is sulfuric acid, as shown by the following equations.

$$H_2SO_4(l) + H_2O(l) \longrightarrow H_3O^+(aq) + HSO_4^-(aq)$$

$$HSO_4^-(aq) + H_2O(l) \longrightarrow H_3O^+(aq) + SO_4^{2-}(aq)$$

As shown in the above equations, sulfuric acid has two hydrogen atoms that can form hydronium ions in solution. Therefore, H_2SO_4 produces more electrolytes in solution than does HCl.

Self-Check How many hydronium ions can H_3PO_4 form in solution?

An **Arrhenius base** is defined as any substance that produces hydroxide, OH^-, ions when it is added to water. For example, NaOH is a base as shown by the following equation.

$$NaOH(s) \longrightarrow Na^+(aq) + OH^-(aq)$$

A strong base, such as NaOH, dissociates in water to produce a high concentration of OH^- ions. In contrast, a weak base, such as NH_3, produces relatively few OH^- ions.

Self-Check Examine the following reaction.
$K_2O(s) + H_2O(l) \longrightarrow 2K^+(aq) + 2OH^-(aq)$
Is $K_2O(s)$ an acid or a base? Explain your answer.

The Arrhenius definitions of an acid and base cannot be applied in all cases. For example, some compounds can act either as an Arrhenius acid or as an Arrhenius base, depending on the substance with which it is reacting. As a result, another definition of acids and bases was proposed in 1923. This definition states that an **acid** is a substance that donates a proton to another substance. A proton is a hydrogen atom that has lost its electron. For example, consider the following reaction.

$$HCl(g) + H_2O(l) \longrightarrow H_3O^+(aq) + Cl^-(aq)$$

Notice that HCl donates a proton, H^+, to H_2O. As a proton donor, HCl is considered an acid. *Reference Table K* lists some common acids.

A **base** is defined as a substance that accepts a proton. Take another look at the reaction where HCl is added to water.

$$HCl(g) + H_2O(l) \longrightarrow H_3O^+(aq) + Cl^-(aq)$$
acid   base

Unit XI Acids and Bases continued

Notice that H_2O accepts the proton that HCl donates. Therefore, H_2O acts as a base in the reaction shown above. *Reference Table L* lists some common bases. If you check this table, you will notice that three of the compounds listed contain OH^- ions. The fourth compound, NH_3, does not.

> **Self-Check** Examine the following reaction.
> $NH_3(aq) + H_2O(l) \rightarrow NH_4^+(aq) + OH^-(aq)$
> Identify the acid and the base.

Defining an acid as a proton donor and a base as a proton acceptor can be useful in certain situations. Consider the following reaction.

$$\underset{\text{acid}}{HCO_3^-(aq)} + \underset{\text{base}}{NH_3(aq)} \rightarrow CO_3^{2-}(aq) + NH_4^+(aq)$$

Notice that NH_3 accepts a proton to form NH_4^+. Therefore, NH_3 is a base. However, NH_3 would not be considered a base according to the Arrhenius definition because it does not produce any hydroxide ions in solution.

REVIEW YOUR UNDERSTANDING

_____ 1. Based on the Arrhenius definition, which ion does a base produce in aqueous solution?
   (1) H^+
   (2) H_3O^+
   (3) OH^-
   (4) hydronium ion

_____ 2. A strong acid, such as HCl,
   (1) produces few electrolytes in solution.
   (2) dissociates only partially when added to water.
   (3) does not conduct electricity when in aqueous solution.
   (4) completely dissociates in aqueous solution to form electrolytes.

_____ 3. A substance that accepts a proton is defined as
   (1) an acid.
   (2) a base.
   (3) either an acid or a base.
   (4) a hydronium ion.

Unit XI Acids and Bases continued

Given the following reaction:

$$CH_3COOH(aq) + H_2O(l) \longrightarrow H_3O^+(aq) + CH_3COO^-(aq)$$

_____ 4. In the above reaction, $H_2O(l)$ acts as a(n)
(1) acid.
(2) base.
(3) proton acceptor.
(4) electrolyte.

_____ 5. Which compound donates a proton and is therefore an acid?
(1) $CH_3COOH(aq)$
(2) $H_2O(l)$
(3) $H_3O^+(aq)$
(4) $CH_3COO^-(aq)$

pH and Titration

You learned that strong acids and bases completely dissociate in aqueous solutions. In contrast, weak acids and bases dissociate only partially. However, there is another way to describe the strength of an acid and a base.

This method uses a scale to describe the relative strengths of acids and bases. This is the pH scale. The **pH** scale is based on the concentration of hydronium ions in solution. The scale is actually a negative logarithm because it uses the negative of the power of 10.

The pH scale ranges from 0 to 14, with 7 being the midpoint. A solution that has a pH value of 7 is neutral, which is it is neither acidic nor basic. A solution with a pH value less than 7 is acidic. The lower the pH value is, the more acidic the solution. A solution with a pH value greater than 7 is basic. The higher the PH value is, the more basic the solution.

Self-Check Which is the more acidic solution, one that has a pH of 6 or one that has a pH of 4?

If the concentration of hydronium ions in solution is 1×10^{-6}, then the pH of this solution is 6. If the concentration of hydronium ions in solution is 1×10^{-5}, then the pH is 5. Notice that the difference in pH values between these two solutions is 1. However, pH is based on a power of 10. Therefore, a solution with a pH of 5 has a hydronium ion concentration that is ten times greater than the hydronium concentration in the solution with a pH of 6.

Unit XI Acids and Bases continued

Self-Check The pH of a solution changes from a pH of 3 to a pH of 5. How has the hydronium ion concentration changed?

The pH of a solution can be measured with a pH meter. A less expensive way to determine the pH of a solution is to use an indicator. An indicator is a compound that can reversibly change color depending on the pH of the solution. For example, thymol blue is an **indicator.** When added to solutions whose pH is between 3 and 8, thymol blue is yellow. In solutions whose pH is 10 or higher, thymol blue is blue.

Reference Table M lists some common indicators. This table indicates the pH range over which the indicator changes color. For example, phenolphthalein is colorless is a solution whose pH value is less than 8.2 However, as the pH value increases from 8.2 to 10, phenolphthalein turns pink.

Self-Check A colorless solution has a pH value of 8. What color will this solution turn if methyl orange is added?

Indicators are often used in a laboratory procedure known as a **titration.** A titration is a process in which a volume of solution of known concentration is used to determine the concentration of another solution. For example, a student titrates 40 mL of an HCl solution of unknown concentration with 25 mL of 0.5 M NaOH. The student can use the following equation to calculate the concentration of the HCl solution.

volume of the acid × molarity of the acid = volume of the base × molarity of the base

$$V_A M_A = V_B M_B$$

40 mL HCl × ? M HCl = 25 mL NaOH × 0.5 M NaOH

$$? \text{ M HCl} = \frac{(25 \text{ mL})(0.5 \text{ M})}{50 \text{ mL}} = 0.25 \text{ M}$$

Self-Check If 5.0 mL of a 0.20 M HCl solution is used to titrate 10 mL of NaOH, what is the concentration of the base?

Sometimes a titration is performed using a strong acid, such as HCl, and a strong base, such as NaOH. When a solution of a strong acid and a strong base with exactly equal amounts of $H_3O^+(aq)$ and $OH^-(aq)$ ions, are mixed, almost all these ions react to form water.

Notes/Study Ideas/Answers

Unit XI Acids and Bases continued

This reaction is sometimes written as follows.

$$HCl + NaOH \longrightarrow NaCl + H_2O$$

The reaction shown above is considered a **neutralization reaction** in which an acid (HCl) reacts with a base (NaOH) to produce a salt (NaCl) plus water (H_2O).

REVIEW YOUR UNDERSTANDING

_____ 6. Which pH represents a solution that has the highest concentration of hydronium ions?
  (1) pH 3
  (2) pH 5
  (3) pH 7
  (4) pH 9

_____ 7. What is the pH of a solution that has 100 times the hydronium concentration of a solution whose pH is 9?
  (1) pH 3
  (2) pH 5
  (3) pH 7
  (4) pH 9

_____ 8. What are the two products of a neutralization reaction?
  (1) an acid and a base
  (2) an acid and a salt
  (3) a base and water
  (4) a salt and water

_____ 9. What salt will be formed when HNO_3 reacts with NaOH in a neutralization reaction?
  (1) sodium hydroxide
  (2) sodium sulfate
  (3) sodium nitrate
  (4) sodium nitride

_____ 10. If a solution has a pH value of 7, then the
  (1) hydronium ion concentration is greater than the hydroxide ion concentration.
  (2) hydronium ion concentration is equal to the hydroxide ion concentration.
  (3) hydronium ion concentration is less than the hydroxide ion concentration.
  (4) hydronium ion concentration equals 1×10^{-8} M.

Unit XI Acids and Bases *continued*

_____ **11.** What color is bromcresol green in a basic solution?
  (1) green
  (2) pink
  (3) yellow
  (4) blue

ANSWERS TO SELF-CHECK QUESTIONS

- It can form three hydronium ions.

- It is a base, because $K_2O(s)$ produces OH^- ions in aqueous solution.

- In this reaction H_2O is an acid because it donates a proton to NH_3, which is the base because it accepts the proton.

- The solution with a pH of 4 is more acidic.

- The hydronium ion concentration is 100 times less concentrated compared to the original solution.

- The solution will turn yellow.

- The concentration of the base is 0.10 M.

Questions for Regents Practice

UNIT XI

Holt Chemistry: The Physical Setting

Acids and Bases

PART A

_____ 1. Which ion is produced when an Arrhenius base is dissolved in water?
(1) H^+, as the only positive ion in solution
(2) H_3O^+, as the only positive ion in solution
(3) OH^-, as the only negative ion in solution
(4) H^-, as the only negative ion in solution

_____ 2. Which reaction is an example of a neutralization reaction?
(1) $CH_3COOH(aq) + H_2O(l) \rightarrow H_3O^+(aq) + CH_3COO^-(aq)$
(2) $HNO_3(l) + H_2O(l) \rightarrow H_3O^+(aq) + NO_3^-(aq)$
(3) $HCl(aq) + LiOH(aq) \rightarrow H_2O(l) + LiCl(aq)$
(4) $H_2SO_4(l) + H_2O(l) \rightarrow H_3O^+(aq) + HSO_4^-(aq)$

_____ 3. Which substance is an Arrhenius acid?
(1) HF
(2) KCl
(3) $Mg(OH)_2$
(4) CH_4

_____ 4. Which of these 1 M solutions will have the highest pH value?
(1) HCl
(2) NaCl
(3) CH_3COOH.
(4) NaOH

_____ 5. Which reaction occurs when H_3O^+ ions react with OH^- ions to form H_2O?
(1) titration
(2) neutralization
(3) dissociation
(4) ionization

_____ 6. When the pH of a solution changes from pH 8 to pH 10, the hydronium concentration is
(1) 0.01 of the original content.
(2) 0.1 of the original content.
(3) 10 times the original content.
(4) 100 times the original content.

_____ 7. Which indicator is yellow in basic solutions?
(1) methyl orange
(2) bromothymol blue
(3) litmus
(4) bromcresol green

PART B-1

_____ 8. Which compound produces a water solution that conducts electricity and turns phenolphthalein pink?
(1) NaCl
(2) KOH
(3) H_2SO_4
(4) HCl

_____ 9. Identify the salt that remains when a solution of H_2SO_4 is titrated with a solution of $Ca(OH)_2$.
(1) calcium hydroxide
(2) calcium oxide
(3) calcium sulfate
(4) calcium phosphate

Unit XI Acids and Bases continued

___ 10. Which of the following can be used to titrate a 0.1 M HCl solution?
(1) 0.1 M NaCl
(2) 0.5 M H₂SO₄
(3) 0.1 M KOH
(4) 0.1 M KCl

___ 11. If 25 mL of a 0.20 M NaOH solution is used to titrate 100 mL of HCl, what is the molarity of the acid solution?
(1) 0.05 M
(2) 0.10 M
(3) 0.20 M
(4) 0.80 M

___ 12. A student was given three unknown solutions. Each solution was tested for conductivity and also tested with phenolphthalein. The results are shown in the data table below.

Solution	Conductivity	Color with Phenolpthalein
A	good	colorless
B	good	pink
C	poor	colorless

Which could be a 0.1 M NaOH solution?
(1) A, only
(2) B, only
(3) C, only
(4) A and C, both

PART B-2

13. Explain why strong acids and bases are good electrical conductors.

_____

_____

_____

14. Write the balanced equation that shows the neutralization reaction that occurs when equal concentrations and volumes of HNO₃ and Ca(OH)₂ are mixed.

15. What volume of 0.10 M NaOH is required to neutralize 25 mL of 0.50 M HCl?

16. The table below lists various indicators and the pH transition range for their color change.

Indicator	Acid color	pH transition range	Base color
Thymol blue	red	1.2–2.8	yellow
Bromphenol blue	yellow	3.0–4.6	blue
Bromcresol green	yellow	2.0–5.6	blue
Bromthymol blue	yellow	6.0–7.6	blue
Phenol red	yellow	6.6–8.0	red
Alizarin yellow	yellow	10.1–12.0	red

Select an indicator that would be appropriate to use for a titration involving a strong acid and a strong base. Explain your answer.

_____

_____

Unit XI Acids and Bases continued

_____ 17. Explain how the following equation shows that water has the properties of both an acid and a base.

$$H_2O(l) + H_2O(l) \rightarrow H_3O^+(aq) + OH^-(aq)$$

PART C

A student recorded the following readings during a titration of a base with an acid of known molarity.

	0.10 M HCl	NaOH
Initial reading	12.02 mL	6.02 mL
Final reading	22.03 mL	11.03 mL

18. Calculate the molarity of the NaOH.

19. Explain why it is better to average the data from several titrations rather than the data from a single titration to determine the molarity of an unknown solution.

20. Antacid products are taken to reduce excess stomach acid. What do all antacid products contain as the active ingredient? Explain your answer.

Name _____ Class _____ Date _____

UNIT XII
Entropy and Equilibrium

Holt Chemistry: The Physical Setting

Many chemical reactions are spontaneous. These reactions proceed at room temperature without the addition of a catalyst. Just mixing the reactants is enough to bring about the reaction. A reaction is always spontaneous if two conditions are met. First, the total potential energy content of the products is lower than the total potential energy content of the reactants. In other words, the reaction is exothermic. Second, the products display a greater disorder than the reactants. The measure of the randomness or disorder of a system is known as **entropy**.

Energy and Entropy

The total energy content of a system can be measured. This energy content is represented by the symbol H. When a chemical reaction takes place, there is a change in the value of H. This change is represented by the symbol ΔH, where Δ is the Greek letter *delta* and represents the term *change*.

The ΔH value can be either positive or negative. For example, consider the following reaction.

$$H_2(g) + Br_2(g) \longrightarrow 2HBr(g)$$

For the reaction shown above, $\Delta H = -72.8$ kilojoules (kJ). This ΔH indicates that when 1 mol $H_2(g)$ combines with 1 mol $Br_2(g)$ to form 2 mol $HBr(g)$, 72.8 kJ of energy is released. Because energy is released, this is an exothermic reaction. Exothermic reactions have negative ΔH values. Systems in nature tend toward lower energy. Therefore, a negative ΔH value is an indication that a reaction is likely to be spontaneous.

However, there is another force that affects natural systems, such as physical changes and chemical reactions. This is entropy. A system with greater disorder has greater entropy. Systems in nature tend toward higher entropy. For example, consider the following physical change.

$$H_2O(s) \longrightarrow H_2O(l)$$

The equation above represents a change in state that occurs naturally at room temperature—ice melting to form water. You learned that the melting of a solid is an endothermic process represented by the heat of fusion. Therefore, the ΔH value for ice melting must be positive because this is an endothermic process.

What You'll Need to Learn
This topic is part of the Regents Curriculum for the Physical Setting Exam. Standard 4, Performance Indicator 3.1: *Explain the properties of materials in terms of the arrangement and properties of the atoms that compose them.* Standard 4, Performance Indicator 3.4: *Use kinetic molecular theory (KMT) to explain rates of reactions and the relationships among temperature, pressure, and volume of a substance.*

What Terms You'll Need to Know
chemical equilibrium
entropy
Le Châtelier's principle
phase equilibrium
sublimation
vapor pressure

Where You Can Learn Even More:
Holt Chemistry: The Physical Setting
Chapter 10: Causes of Change
Chapter 11: States of Matter and Intermolecular Forces (Section 4)
Chapter 14: Chemical Equilibrium

Self-Check Does $H_2O(g) \longrightarrow H_2O(l)$ have a positive or negative ΔH value? Explain your answer.

Copyright © by Holt, Rinehart and Winston. All rights reserved.

Entropy and Equilibrium

Unit XII Entropy and Equilibrium continued

Notes/Study Ideas/Answers

Although the melting of ice is an endothermic process, it occurs spontaneously. How can this be if the tendency in nature is toward lower energy? The answer is that entropy, the other driving force in nature, is more significant in this case. Liquids have greater entropy than solids, and gases have greater entropy than liquids. Ice melts spontaneously because the tendency toward greater entropy is more significant than the tendency toward higher energy content, or a positive ΔH value.

Reference Table I lists the ΔH values for various chemical reactions. Any negative ΔH value indicates that the reaction as written is exothermic. Keep in mind that this does not indicate that the reaction is spontaneous because you must also know how the reaction changes with respect to entropy.

REVIEW YOUR UNDERSTANDING

_____ 1. Which of the following is a natural tendency for a system?
   (1) toward lower energy
   (2) toward higher energy
   (3) toward greater order
   (4) toward a higher ΔH value

_____ 2. If a reaction releases energy, then it is
   (1) endothermic with a positive ΔH value.
   (2) endothermic with a negative ΔH value.
   (3) exothermic with a positive ΔH value.
   (4) exothermic with a negative ΔH value.

_____ 3. Entropy refers to
   (1) the energy content of a system.
   (2) the degree of disorder of a system.
   (3) endothermic reactions.
   (4) exothermic reactions.

_____ 4. Which substance has the highest degree of entropy?
   (1) $H_2O(s)$
   (2) $H_2O(l)$
   (3) $H_2O(g)$
   (4) $CH_3COOH(aq)$

_____ 5. Which reaction is endothermic?
   (1) $H_2O(l) \rightarrow H_2O(s)$
   (2) $C(s) + O_2(g) \rightarrow CO_2(g)$
   (3) $NaCl(s) \rightarrow Na^+(aq) + Cl^-(aq)$
   (4) $H^+(aq) + OH^-(aq) \rightarrow H_2O(l)$

Copyright © by Holt, Rinehart and Winston. All rights reserved.
Entropy and Equilibrium

Unit XII Entropy and Equilibrium continued

Equilibrium

Place ice cubes in a glass of water at room temperature, and they will eventually melt. In this case, the phase change proceeds in only one direction from solid to liquid. However, place ice cubes in a glass of water at 0°C, and they will not melt. The solid ice and liquid water are in **phase equilibrium**.

At equilibrium, the H_2O molecules are constantly moving between the solid and liquid phases. Molecules are moving from the solid phase to the liquid phase at the same rate that other molecules are moving from the liquid phase to the solid phase. In other words, the rate of melting equals the rate of freezing.

Phase equilibrium can also exist between a liquid and a gas. An example is a closed container of water. Molecules of water are in a gas phase in the space just above the liquid phase. At equilibrium, the rate of evaporation equals the rate of condensation.

At equilibrium, the gas (vapor) molecules exert a pressure on the liquid molecules. This pressure is called the **vapor pressure**. The vapor pressure depends on the temperature. As the temperature increases, the water molecules have more kinetic energy. As a result, more of them escape into the gas phase. With more molecules escaping into the gas phase, the vapor pressure increases. *Reference Table H* shows the relationship between temperature and vapor pressure of four liquids.

Self-Check What is the vapor pressure of ethanol at 90°C?

Phase equilibrium can also exist between a solid and a gas. The process by which a solid turns directly into a gas is called sublimation. The following equation is an example of **sublimation**. Notice that the arrows point in both the forward and reverse direction to indicate that this process is in equilibrium.

$$CO_2(s) \rightleftarrows O_2(g)$$

Many reactions proceed in only one direction. This means that when the reaction is over, only a trace of the reactants remains. In effect, almost all their atoms have been rearranged to form products. Such reactions are called completion reactions.

In contrast, other reactions are called reversible reactions because the products can reform reactants. Arrows that point in both directions are used to show that a reaction is reversible. For example, consider the following reaction.

$$N_2(g) + 3H_2(g) \rightleftarrows 2NH_3(g)$$

Unit XII Entropy and Equilibrium *continued*

Self-Check The ΔH value for this reaction is –91.8 kJ. Should this reaction proceed spontaneously based on this change in energy content?

In the reaction shown above, nitrogen and hydrogen combine to form ammonia. However, the ammonia also decomposes to reform the reactants. The reaction shown above, therefore, reaches a **chemical equilibrium**.

A chemical equilibrium means that the rate of a forward reaction equals the rate of the reverse reaction. At equilibrium, the concentrations of products and reactants remain unchanged. However, this does not mean that the concentrations of reactants and products are equal. For example, if the concentrations of the reactants and product in the reaction shown above were measured, the data would show that the concentration of $NH_3(g)$ is much greater than the concentrations of both $H_2(g)$ and $N_2(g)$. However, at equilibrium, when 1 mol $N_2(g)$ and 3 mol $H_2(g)$ combine to form 2 mol $NH_3(g)$, at the same time, 2 mol $NH_3(g)$ decompose to reform 1 mol $N_2(g)$ and 3 mol $H_2(g)$. Therefore, at equilibrium, the concentrations of reactants and products remain constant.

REVIEW YOUR UNDERSTANDING

_____ **6.** An arrow in an equation that points in only the forward direction indicates that the reaction
   (1) goes to completion.
   (2) is reversible.
   (3) is in equilibrium.
   (4) shows that the products have less energy that the reactants.

_____ **7.** At equilibrium, the concentrations of the reactants and the products
   (1) are equal.
   (2) cannot be measured.
   (3) remain constant.
   (4) are equal to the concentrations present at the start of the reaction.

_____ **8.** The change of a solid directly into a gas is an example of
   (1) a phase equilibrium.
   (2) an exothermic process.
   (3) a chemical equilibrium.
   (4) sublimation.

_____ 9. Which exerts the lowest vapor pressure at 50°C?
(1) propanone
(2) ethanol
(3) water
(4) ethanoic acid

Le Châtelier's Principle

A chemist named Henri Le Châtelier studied how reactions at equilibrium responded to changes, such as decreasing the temperature or increasing the pressure. His studies led to **Le Châtelier's principle**, which states that a system in equilibrium will oppose a change in a way that helps eliminate the change.

Take another look at a reaction you studied earlier.

$$N_2(g) + 3H_2(g) \rightleftharpoons 2NH_3(g) + 92 \text{ kJ}$$

This reaction is the Haber process, which is used to synthesize ammonia for a variety of commercial uses. Notice that the Haber process is exothermic. Do not become confused by the portion of the equation that shows "+ 92 kJ." This simply shows that the energy released as heat can be considered a product. In this case, the + symbol in the equation represents *plus* and not a positive sign. The ΔH value for this exothermic reaction would be written as –92 kJ.

Changing conditions places a stress on reactions in equilibrium. According to Le Châtelier's principle, the system will react to eliminate this stress.

Recall that the Haber process is an exothermic reaction. Therefore, an increase in temperature will place a stress on the right side of the equation where energy is shown as a product. To relieve this stress, the equilibrium will be shifted to the left. In other words, the reverse reaction will be favored.

This means that the concentration of $NH_3(g)$ will decrease and the concentrations of $N_2(g)$ and $H_2(g)$ will increase until a new equilibrium is established. Once the new equilibrium is established, the concentrations of reactants and product will remain constant.

Self-Check How will the Haber process at equilibrium be affected if the temperature is lowered?

Reactions in equilibrium that involves gases are also affected by changes in pressure. Once again, take a look at the Haber process.

$$N_2(g) + 3H_2(g) \rightleftharpoons 2NH_3(g)$$

Unit XII Entropy and Equilibrium *continued*

Notice that the forward reaction converts 4 mol of $N_2(g)$ and $H_2(g)$ into 2 mol of $NH_3(g)$. In other words, the product takes up less volume. You learned that as the pressure is increased, the volume of a gas decreases. Therefore, increasing the pressure on a system at equilibrium favors the direction that leads to the smaller volume of gases. In the Haber process, an increase in pressure, therefore, favors the forward reaction. In contrast, lowering the pressure favors the reverse reaction.

Changing the concentrations of the substances involved in a reaction also affects the equilibrium. For example, adding more $N_2(g)$ to the reaction mixture places stress on the left side of the equation and therefore favors the forward reaction. Increasing the concentration of $H_2(g)$ will also favor the forward reaction to reduce the stress placed on the left side of the equation

REVIEW YOUR UNDERSTANDING

_____ 10. If an exothermic reaction has reached equilibrium, then increasing the temperature will
   (1) favor the forward reaction.
   (2) favor the reverse reaction.
   (3) favor both the forward and reverse reactions.
   (4) not affect the equilibrium.

_____ 11. Le Châtelier's principle states that
   (1) at equilibrium, the rate of the forward reaction is equal to the rate of the reverse reaction.
   (2) changes in pressure affect equilibrium reactions that involve only solids.
   (3) changes in temperature affect only equilibrium reactions that are exothermic.
   (4) chemical equilibrium reactions respond to reduce applied stress.

Examine the following equilibrium reaction.

$$2NO_2(g) \rightleftarrows N_2O_4(g) + 55.3 \text{ kJ}$$

_____ 12. Increasing the temperature will favor
   (1) the forward reaction.
   (2) the reverse reaction.
   (3) an increase in the concentration of $N_2O_4(g)$.
   (4) a decrease in the concentration of $NO_2(g)$.

Unit XII Entropy and Equilibrium *continued*

_____ 13. Decreasing the pressure will favor
  (1) the forward reaction.
  (2) the reverse reaction.
  (3) an increase in the concentration of $N_2O_4(g)$.
  (4) a decrease in the concentration of $NO_2(g)$.

ANSWERS TO SELF-CHECK QUESTIONS

- This process has a negative ΔH value because the condensation of water vapor is an exothermic process.

- 150 kPa

- The reaction should proceed spontaneously because it is exothermic. However, the change in entropy must also be taken into consideration, along with the temperature at which the reaction is taking place.

- The forward reaction will be favored at low temperatures.

Questions for Regents Practice

UNIT XII

Holt Chemistry: The Physical Setting

Part A

_____ 1. Given the reaction:
$$CH_4(g) + 2O_2(g) \rightarrow 2H_2O(g) + CO_2(g)$$

What is the overall result when $CH_4(g)$ burns according to this reaction?
(1) Energy is absorbed and ΔH is negative.
(2) Energy is absorbed and ΔH is positive.
(3) Energy is released and ΔH is negative.
(4) Energy is released and ΔH is positive.

_____ 2. The measure of the randomness of a system is known as its
(1) phase equilibrium.
(2) ΔH value.
(3) entropy.
(4) sublimation temperature.

_____ 3. 3. Which represents a decrease in entropy?
(1) $H_2O(l) \rightarrow H_2O(g)$
(2) $NaCl(s) \rightarrow Na^+(aq) + Cl^-(aq)$
(3) $CO_2(s) \rightarrow CO_2(g)$
(4) $2Na(g) + Cl_2(g) \rightarrow 2NaCl(s)$

_____ 4. The liquid and gas phases of water can exist in a state of equilibrium at 1 atmosphere of pressure and a temperature of
(1) 0°C
(2) 100°C
(3) 273°C
(4) 100 K

_____ 5. At equilibrium, the concentrations of the reactants and the products
(1) remain constant.
(2) are equal.
(3) decrease.
(4) increase.

_____ 6. Which type of change, if any, can reach equilibrium?
(1) a chemical change, only
(2) a physical change, only
(3) both a physical change and a chemical change
(4) neither a physical change nor a chemical change

_____ 7. Given the reaction:
$$AgCl(s) \rightleftarrows Ag^+(aq) + Cl^-(aq)$$

Which statement is true once this reaction reaches equilibrium?
(1) The entropy of the forward reaction will continue to increase.
(2) The $AgCl(s)$ will be completely consumed.
(3) The concentration of $Ag^+(aq)$ will be greater than the concentration of $Cl^-(aq)$.
(4) $AgCl(s)$ will dissociate at the same rate as it forms.

_____ 8. According to Le Châtelier's principle, which change will place stress upon a system?
(1) increasing the temperature
(2) adding a catalyst to a reaction
(3) lowering the energy of activation
(4) maintaining the pressure at 1 atmosphere and the temperature at 0°C

Entropy and Equilibrium

Unit XII Entropy and Equilibrium continued

PART B-1

9. The vapor pressure of propanone at 62°C is 120 kPa. At what temperature does water have this same vapor pressure?
   (1) 62°C
   (2) 100°C
   (3) 105°C
   (4) 125°C

Given the equilibrium reaction at 101.3 kPa and 298 K:

$$2C(s) + O_2(g) \rightleftarrows 2CO(g) + 566 \text{ kJ}$$

10. Which will favor the forward reaction?
    (1) increasing the pressure
    (2) decreasing the pressure
    (3) adding a catalyst
    (4) increasing the temperature

11. How will the reaction shown above be affected if the temperature is changed to 60°C?
    (1) The concentration of $O_2(g)$ will decrease.
    (2) The concentration of $C(s)$ will decrease.
    (3) The concentration of $O_2(g)$ will increase.
    (4) The forward reaction will be favored.

12. Changes in pressure affect chemical reactions at equilibrium if they involve
    (1) solids.
    (2) liquids.
    (3) gases.
    (4) ions in aqueous solution.

Given the equilibrium reaction in a closed system:

$$H_2(g) + I_2(g) + \text{heat} \rightleftarrows 2HI(g)$$

13. How will a temperature increase affect this reaction?
    (1) The equilibrium will shift to the right and the concentration of HI will decrease.
    (2) The equilibrium will shift to the right and the concentrations of H_2 and I_2 will increase.
    (3) The equilibrium will shift to the left and the concentration of H_2 will decrease.
    (4) The equilibrium will shift to the left and the concentration of HI will increase.

14. In the reaction shown above, an increase in pressure will favor
    (1) the forward reaction.
    (2) the reverse reaction.
    (3) neither the forward nor the reverse reaction.
    (4) the tendency for this reaction to go to completion.

Unit XII Entropy and Equilibrium continued

PART B-2

Given the reaction at equilibrium:

$$2A(g) + B(g) \rightleftarrows A_2B(g) + \text{heat}$$

15. Discuss two ways to increase the concentration of $A_2B(g)$ produced in this reaction.

16. Explain why a saturated solution is considered to be in equilibrium.

17. Explain why water evaporating from a glass is not an equilibrium system.

PART C

Consider the following graph that plots reaction rate over time for the following reaction.

$$H_2(g) + I_2(g) \rightleftarrows 2HI(g)$$

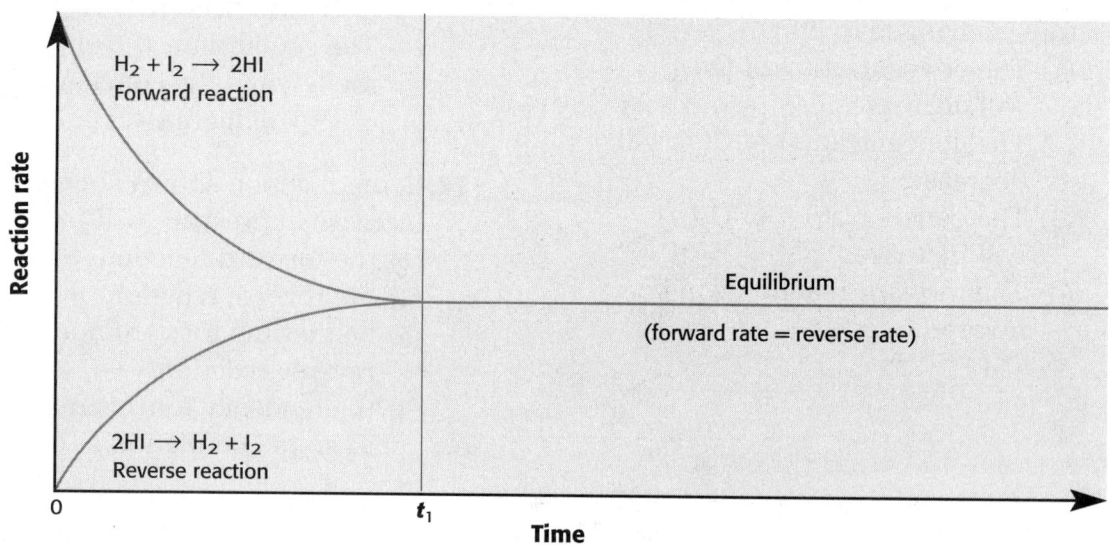

Unit XII Entropy and Equilibrium *continued*

18. Explain what is happening to the rate of the forward reaction between the time values 0 and t_1.

19. Explain what is happening to the rate of the reverse reaction between the time values 0 and t_1.

20. How will a change in pressure affect the value of the reaction rate at equilibrium?

UNIT XIII
Organic Chemistry

Holt Chemistry: The Physical Setting

As an element, carbon can form various bonding arrangements, known as *allotropes*. For example carbon atoms can bond to one another to form diamond, which is the hardest known substance. Carbon atoms can also bond to one another to form graphite, a substance whose properties make it useful as a lubricant and as pencil lead. Other substances that are formed by the bonding of carbon atoms include fullerenes, which consist of interconnecting rings, and nanotubes, which are hollow cylinders about 10,000 times narrower than a human hair.

Hydrocarbons

Most compounds of carbon are known as **organic compounds**. An organic compound consists of carbon atoms, of course, and most also contain hydrogen atoms. The simplest class of organic compounds consists of **hydrocarbons**. A hydrocarbon is an organic compound composed of only carbon and hydrogen.

The simplest hydrocarbons are the **alkanes**. An alkane is a hydrocarbon whose carbon atoms are connected only by single bonds. The three simplest hydrocarbons are methane, ethane, and propane. The structural formulas for each of these hydrocarbons are drawn as follows.

methane, CH_4 ethane, C_2H_6 propane, C_3H_8

Notice that these structural formulas differ from one another by one carbon atom and two hydrogen atoms. Therefore, the general formula for an alkane can be written as C_nH_{2n+2}, where n represents the number of carbon atoms. For example, the alkane that contains 15 carbon atoms has the molecular formula $C_{15}H_{32}$.

Another class of hydrocarbons consists of the **alkenes**. An alkene is a hydrocarbon that has at least one double bond between two carbon atoms. The structural formulas for two alkenes are drawn as follows.

ethene   propene

What You'll Need to Learn

This topic is part of the Regents Curriculum for the Physical Setting Exam.
Standard 4, Performance Indicator 3.1: *Explain the properties of materials in terms of the arrangement and properties of the atoms that compose them.*
Standard 4, Performance Indicator 3.2: *Use atomic and molecular models to explain common chemical reactions.*

What Terms You'll Need to Know

addition reaction
alkane
alkene
alkyne
combustion reaction
esterification reaction
fermentation reaction
functional group
hydrocarbon
isomer
organic compound
polymerization reaction
saponification reaction
saturated hydrocarbon
substitution reaction
unsaturated hydrocarbon

Where You Can Learn Even More:

Holt Chemistry: The Physical Setting
Chapter 19: Carbon and Organic Compounds

Unit XIII Organic Chemistry continued

With twice as many hydrogen atoms as carbon atoms, an alkene has the general formula C_nH_{2n}. For example, the alkene that contains 15 carbon atoms has the molecular formula $C_{15}H_{30}$.

Self-Check What are the molecular formulas for the alkane and alkene that contain 24 carbon atoms?

A third class of hydrocarbons consists of the **alkynes**. An alkyne contains one or more triple bonds between carbon atoms. The general formula for an alkyne is C_nH_{2n-2}. The structural formula for alkyne is drawn as follows.

$$H-C\equiv C-H$$
ethyne

Self-Check Is $C_{17}H_{34}$ an alkene or an alkyne? Explain your answer.

You may have noticed that the names of three hydrocarbons you have examined are ethane, ethane, and ethyne. The suffix indicates whether the hydrocarbon is an alkane, alkene, or alkyne. The prefix indicates how many carbon atoms are present in the compound. For example, the prefix *eth-* indicates that the compound contains two carbon atoms.

Reference Table P lists the prefixes used for organic compounds that contain from one to ten carbon atoms. *Reference Table Q* lists the three major classes of hydrocarbons, their general formulas, and the structural formulas for ethane, ethane, and ethyne.

Self-Check What are the molecular formulas for hexene and octyne?

Alkanes are considered **saturated hydrocarbons** because all the bonds between the carbon atoms are single bonds. In contrast, the alkenes and the alkynes are considered **unsaturated hydrocarbons** because they contain at least one multiple bond between their carbon atoms.

Hydrocarbons, like many other types of organic compounds, can exist as **isomers**. Isomers are compounds that have the same chemical composition but different structural formulas.

Unit XIII Organic Chemistry continued

For example, examine the following structural formulas for two alkanes.

butane   cyclobutane

If you add the number of carbon atoms and hydrogen atoms in each alkane, you will find that each compound contains four carbon atoms and ten hydrogen atoms. The molecular formula for both alkanes is C_4H_{10}.

However, the atoms are arranged differently in the two compounds. Butane has a linear structure, while cyclobutane has a ring structure. The prefix *cyclo-* indicates that the compound has a ring structure.

Butane and cyclobutane are isomers because they have the same molecular formulas but different structural formulas. Because they have a different arrangement of their atoms, isomers do not have the same properties.

REVIEW YOUR UNDERSTANDING

_____ 1. Which is an alkane?
(1) C_2H_2
(2) C_4H_8
(3) C_8H_{18}
(4) $C_{10}H_{18}$

_____ 2. A saturated hydrocarbon contains
(1) only single bonds between its carbon atoms.
(2) at least one multiple bond between its carbon atoms.
(3) more carbon atoms than hydrogen atoms.
(4) carbon atoms and hydrogen atoms in a 1:2 ratio.

_____ 3. Which compound represents an alkyne?
(1) $C_{14}H_{28}$
(2) $C_{16}H_{34}$
(3) $C_{20}H_{20}$
(4) $C_{22}H_{42}$

Unit XIII Organic Chemistry continued

_____ 4. Isomers have
   (1) the same molecular formula and the same structural formula.
   (2) the same molecular formula but different structural formulas.
   (3) different molecular formulas and different structural formulas.
   (4) the same properties.

_____ 5. Diamond and graphite represent
   (1) alkanes.
   (2) organic compounds.
   (3) allotropes.
   (4) unsaturated hydrocarbons.

Functional Groups

Hydrocarbons are only one class of organic compounds. They are also the only organic compounds that consist solely of carbon and hydrogen. The other classes of organic compounds contain other elements, such as oxygen, nitrogen, sulfur, phosphorus, and the halogens.

These other classes of organic compounds are classified based on the **functional group** they contain. A functional group is a group of atoms that determines the properties of the compound. Functional groups often determine the physical and chemical properties of an organic compound.

For example, consider how the presence of the functional group —OH affects the properties of a compound. An organic compound that contains the —OH functional group is classified as an alcohol. The structural formulas shown below represent butane, an alkane, and 1-butanol, an alcohol. The 1– indicates that the functional group is attached to a C atom at the end.

$$\text{H-C-C-C-C-H} \quad \text{(butane)} \qquad \text{HO-C-C-C-C-H} \quad \text{(1-butanol)}$$

Notice that 1-butanol differs from butane by the presence of the —OH functional group. Butane, which is a gas at room temperature, melts at –138°C and boils at –0.5°C. In contrast, butanol, which is a liquid at room temperature, has much higher melting and boiling points.

Reference Table R lists the functional groups and the names of the classes of organic compounds they form. For example, the functional group —O— is found in a class of organic com-

Unit XIII Organic Chemistry *continued*

pounds called ethers. Notice that the general formulas in *Reference Table R* contain the label *R*. The *R* represents a bonded atom, such as H, or a group of atoms, such as CH_3.

Self-Check To what class of organic compounds does the following compound belong?

$$H-\underset{\underset{H}{|}}{\overset{\overset{H}{|}}{C}}-\underset{\underset{H}{|}}{\overset{\overset{H}{|}}{C}}-\underset{\underset{H}{|}}{\overset{\overset{H}{|}}{C}}-\overset{\overset{O}{\|}}{C}-O-\underset{\underset{H}{|}}{\overset{\overset{H}{|}}{C}}-\underset{\underset{H}{|}}{\overset{\overset{H}{|}}{C}}-H$$

The name of an organic compound is often a clue as to how to draw its structural formula. For example, consider the compound propanoic acid. The prefix *pro-* indicates that this organic compound contains three carbon atoms. The word *acid* indicates that this compound contains the functional group that makes it an organic acid. If you check *Reference Table R*, you will see that the following functional group is present in organic acids.

$$-\overset{\overset{O}{\|}}{C}-OH$$

To draw the structural formula for propanoic acid, begin by drawing the three-carbon backbone.

$$C - C - C$$

Next, add the functional group for an organic acid.

$$C-C-\overset{\overset{O}{\|}}{C}-OH$$

Finally, add hydrogen atoms so that each carbon atom has four bonds, which is the maximum number that it can form.

$$H-\underset{\underset{H}{|}}{\overset{\overset{H}{|}}{C}}-\underset{\underset{H}{|}}{\overset{\overset{H}{|}}{C}}-\overset{\overset{O}{\|}}{C}-OH$$

The molecular formula for propanoic acid is written as CH_3CH_2COOH. Notice that the molecular formula also provides a clue as to how the structural formula is drawn.

Notes/Study Ideas/Answers

Organic Chemistry — Focus On Regents Exam

Unit XIII Organic Chemistry continued

REVIEW YOUR UNDERSTANDING

_____ 6. Which class of organic compounds can be represented by the formula R—O—R?
(1) alcohol
(2) ester
(3) ketone
(4) ether

_____ 7. Which is the functional group for an aldehyde?
(1) O
     ‖
    —C—

(2) O
     ‖
    —C—H

(3) —OH

(4) O
     ‖
    —C—O—

_____ 8. Which compound can be classified as an organic acid?
(1) CH_3OH
(2) $CH_3CH_2COOC_2H_5$
(3) CH_3Cl
(4) CH_3CH_2COOH

_____ 9. Which organic compound is a halide (halocarbon)?
(1) C_2H_4
(2) CH_3COOH
(3) CH_3ClCH_3
(4) $CH_3CH_2OCH_3$

Types of Reactions

Classifying organic reactions helps to see similarities and differences between them. One class of organic reactions is composed of **addition reactions**. An addition reaction involves the addition of an atom or a molecule to an unsaturated molecule and therefore increases the saturation of that molecule.

A common type of addition reaction is hydrogenation, in which one or more hydrogen atoms are added to an unsaturated molecule. Hydrogenation is used to turn vegetable oil into margarine, as shown in the following addition reaction. The brackets around the structural formulas indicate that only a small portion of the very long oil molecule is shown.

Unit XIII Organic Chemistry continued

$$\left(\begin{array}{c}H \\ | \\ -C- \\ | \\ H\end{array} \begin{array}{c}H \\ | \\ C= \end{array} \begin{array}{c}H \\ | \\ C- \end{array} \begin{array}{c}H \\ | \\ C- \\ | \\ H\end{array} \begin{array}{c}H \\ | \\ C= \end{array} \begin{array}{c}H \\ | \\ C- \end{array} \begin{array}{c}H \\ | \\ -C- \\ | \\ H\end{array}\right) + H_2 \xrightarrow{\text{catalyst}} \left(\begin{array}{c}H \\ | \\ -C- \\ | \\ H\end{array} \begin{array}{c}H \\ | \\ C= \end{array} \begin{array}{c}H \\ | \\ C- \end{array} \begin{array}{c}H \\ | \\ C- \\ | \\ H\end{array} \begin{array}{c}H \\ | \\ C- \\ | \\ H\end{array} \begin{array}{c}H \\ | \\ C- \\ | \\ H\end{array} \begin{array}{c}H \\ | \\ -C- \\ | \\ H\end{array}\right)$$

oil   fat

Notice that in the above reaction, one of the double bonds is broken, making room for the addition of two H atoms. This process continues, as two H atoms are added when each double bond between two C atoms in the oil molecule is broken.

Self-Check Explain why alkanes cannot be involved in an addition reaction.

Another type of organic reaction is a **substitution reaction**. A substitution reaction is a reaction in which one or more atoms replace another atom or group of atoms in a molecule.

Alkanes have the lowest chemical reactivity because they are saturated hydrocarbons. However, alkanes can participate in a substitution reaction. Alkanes generally react with a halogen, such as chlorine. As a result of the substitution reaction, a halogen atom replaces a hydrogen atom on the alkane. For example, examine the following substitution reaction between methane and chlorine.

$$\begin{array}{c}H \\ | \\ H-C-H \\ | \\ H\end{array} + Cl-Cl \longrightarrow \begin{array}{c}H \\ | \\ H-C-Cl \\ | \\ H\end{array} + H-Cl$$

methane   chlorine   chloromethane   hydrogen chloride

Catalysts are added to the reaction mixture because single bonds require a large quantity of energy to break. The substitution reaction shown above can continue so that additional chlorine atoms replace the remaining hydrogen atoms.

Self-Check What is the molecular formula for the compound formed when all the hydrogen atoms on methane have been replaced by chlorine atoms in substitution reactions?

A third type of organic reaction includes **polymerization reactions**. A polymerization reaction is similar to an addition reaction. However, a polymerization reaction involves joining smaller molecules to form one, large molecule.

Notes/Study Ideas/Answers

Organic Chemistry   Focus On Regents Exam

Unit XIII Organic Chemistry *continued*

An example can be seen when two ethene, C_2H_4, molecules are joined, as shown in the following polymerization reaction.

$$CH_2{=}CH_2 + CH_2{=}CH_2 \longrightarrow -CH_2-CH_2-CH_2-CH_2-$$

Notice that the product in the above reaction has open bonds at both ends. This indicates that additional ethene molecules can continue to be added to produce a very large molecule called a *polymer*.

Another type of organic reaction is called an **esterification reaction**. This type of reaction produces an ester by reacting an organic acid with an alcohol. For example, ethanoic acid can react with methanol to produce an ester called methyl ethanoate and water.

$$CH_3CH_2OH + CH_3OO \longrightarrow CH_3CH_2OOCH_3 + H_2O$$

A **saponification reaction** involves an ester reacting with a strong base to form an organic acid and an alcohol. The organic acid can then form soap.

A **fermentation reaction** is still another type of organic reaction. Fermentation reactions involve the breakdown of sugars to form alcohol and carbon dioxide. The following reaction shows the fermentation of glucose.

$$C_6H_{12}O_6 \longrightarrow 2C_2H_5OH + 2CO_2$$
$$\text{glucose} \quad \text{ethanol} \quad \text{carbon dioxide}$$

The final type of organic reaction is a **combustion reaction**, in which a compound burns in oxygen to produce carbon dioxide and water. The following equation shows the combustion of propane.

$$C_3H_8 + 5O_2 \longrightarrow 3CO_2 + 4H_2O$$

Unit XIII Organic Chemistry continued

When enough oxygen is not available, the combustion reaction is incomplete. In addition to CO_2 and H_2O, CO and C (unburned carbon) are produced.

REVIEW YOUR UNDERSTANDING

_____ 10. During which type of organic reaction is soap produced?
  (1) polymerization
  (2) fermentation
  (3) saponification
  (4) esterification

_____ 11. Which hydrocarbon *cannot* be involved in an addition reaction?
  (1) propane
  (2) propene
  (3) propyne
  (4) ethene

_____ 12. What are the two products of a fermentation reaction?
  (1) sugar and carbon dioxide
  (2) sugar and alcohol
  (3) alcohol and water
  (4) alcohol and carbon dioxide

Examine the following reaction.

$$C_4H_{10} + Br_2 \longrightarrow C_4H_9Br + HBr$$

_____ 13. The reaction shown above is an example of a(n)
  (1) addition reaction.
  (2) substitution reaction.
  (3) polymerization reaction.
  (4) combustion reaction.

_____ 14. The above equation shows an organic reaction that occurs between an
  (1) alkane and an alkene.
  (2) alkene and a halogen.
  (3) alkyne and a halogen.
  (4) alkane and a halogen.

Notes/Study Ideas/Answers

ANSWERS TO SELF-CHECK QUESTIONS

- alkane: $C_{24}H_{50}$; alkene: $C_{24}H_{48}$

- It is an alkene because it fits the general formula C_nH_{2n}.

- hexene: C_6H_{12}; octyne: C_8H_{14}

- ester

- Only single bonds are present between the carbon atoms in an alkane. Therefore, no multiple bonds are available to be broken.

- CCl_4

- $-CH_2-CH2-CH2-CH2-CH2-CH2-$

- ethanoic acid:
  $$-\overset{\displaystyle O}{\underset{\|}{C}}-OH$$

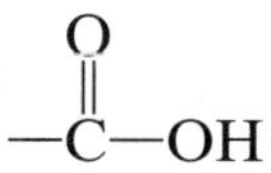

  methanol: $-OH$

  methyl ethanoate:
  $$-\overset{\displaystyle O}{\underset{\|}{C}}-O-$$

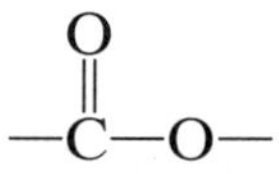

- $C_2H_5OH + 3O_2 \rightarrow 2CO_2 + 3H_2O$

UNIT XIII
Questions for Regents Practice

PART A

1. Given the reaction:

   $CH_4(g) + 2O_2(g) \rightarrow 2H_2O(g) + CO_2(g)$

   What type of organic reaction is shown by the above equation?
   (1) substitution
   (2) addition
   (3) combustion
   (4) fermentation

2. Which formula is an isomer of butane?
   (1) 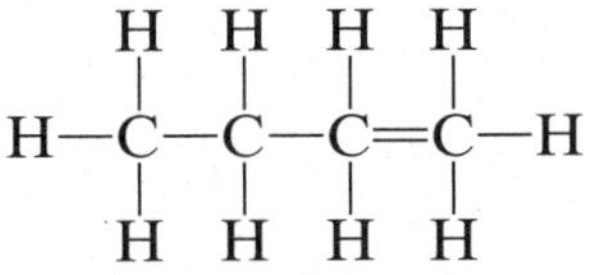 H−C(H)(H)−C(H)(H)−C(H)=C(H)−H
   (2) 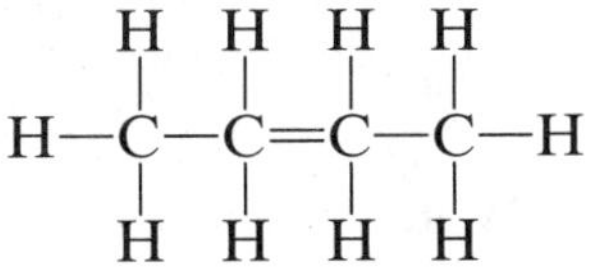 H−C(H)(H)−C(H)=C(H)−C(H)(H)−H
   (3) 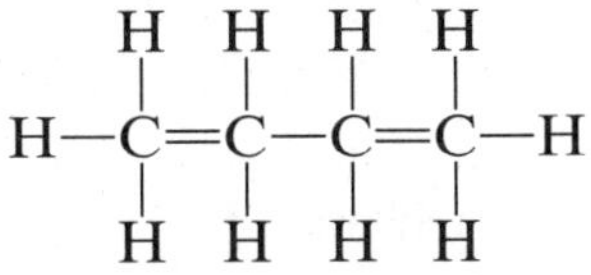 H−C(H)=C(H)−C(H)(H)−C(H)=C(H)−H
   (4) 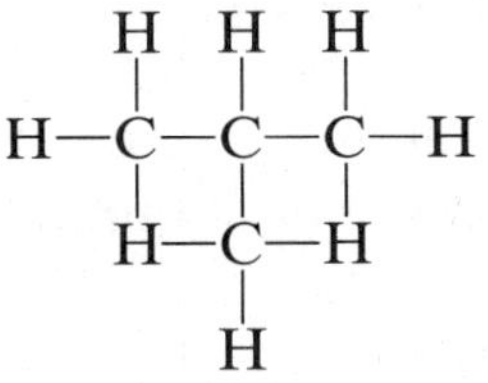 H−C(H)(H)−C(H)(−C(H)(H)(H))−C(H)(H)−H

3. Which compound is classified as a hydrocarbon?
   (1) bromopropane
   (2) propyne
   (3) pronanpol
   (4) diethylether

4. Which compound is an alcohol?
   (1) propanol
   (2) propanal
   (3) propane
   (4) butanoic acid

5. Identify the compound that has an isomer.
   (1) 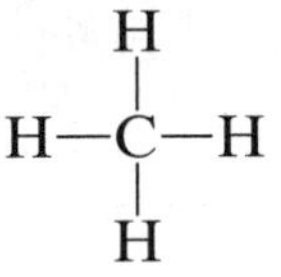 CH₄ structure
   (2) 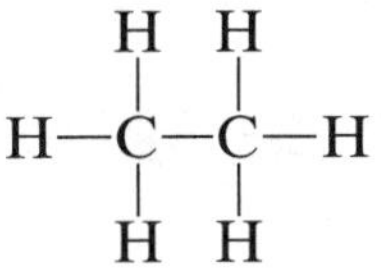 C₂H₆ structure
   (3) 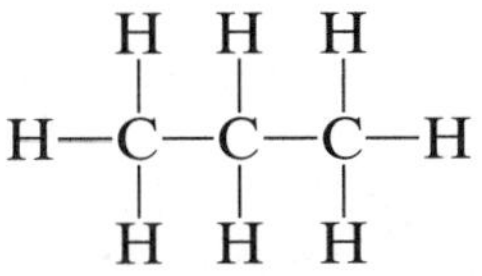 C₃H₈ structure
   (4) 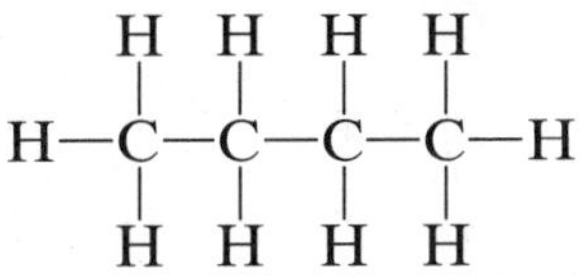 C₄H₁₀ structure

6. In an unsaturated hydrocarbon, carbon atoms are bonded to each other by
   (1) single covalent bonds, only.
   (2) double covalent bonds, only.
   (3) triple covalent bonds, only.
   (4) either one or more multiple covalent bonds.

7. Which formula represents the product of an addition reaction between ethene and bromine?
   (1) CH_3Br
   (2) CH_2Br_2
   (3) C_2H_3Br
   (4) $C_2H_4Br_2$

8. Which compound is an isomer of pentane?
   (1) methyl butane
   (2) butene
   (3) pentyne
   (4) pentanol

Organic Chemistry

Unit XIII Organic Chemistry continued

PART B-1

____ 9. Which pair of compounds are alcohols?

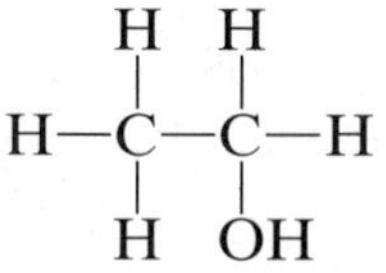

(1)
H H   H H H
H-C-C-H and H-C-C-C-OH
H OH   H H H

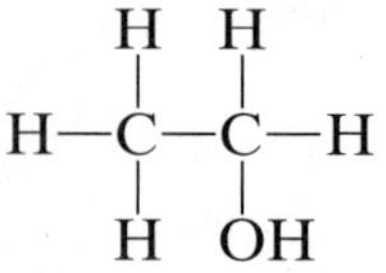

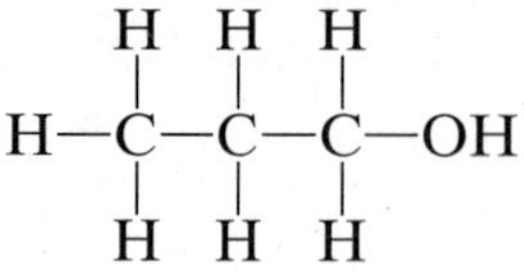

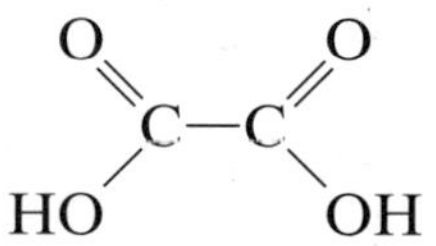

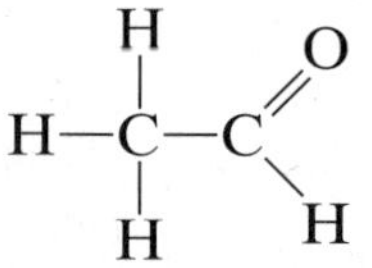

(2)
H H   H   O
H-C-C-H and H-C-C
H OH   H   H

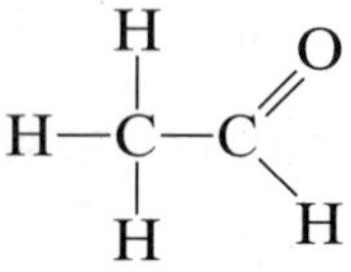

(3)
  O   O   H H H
  C-C   and H-C-C-C-H
HO   OH   H OH H

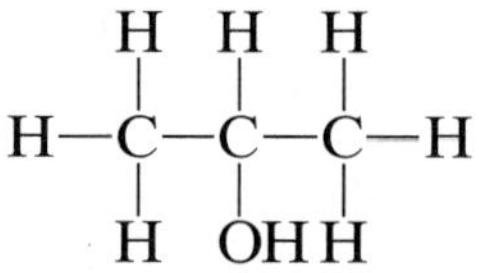

(4)
H   O   H   O
H-C-C   and H-C-C
H   H   H   OH

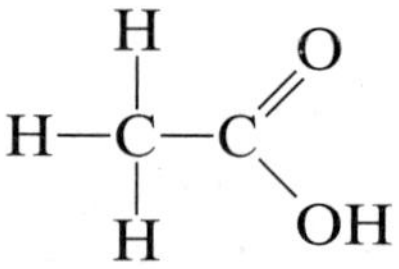

____ 10. Which structural formula correctly represents an unsaturated hydrocarbon?

(1) 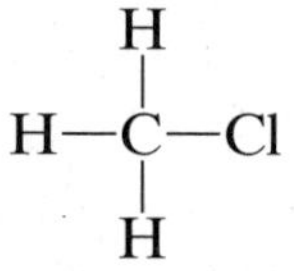
    H
H-C-Cl
    H

(2) 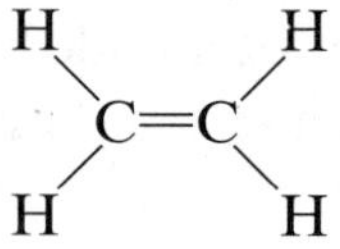
H   H
  C=C
H   H

(3) 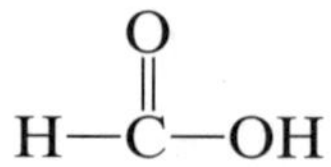
    O
H-C-OH

(4) 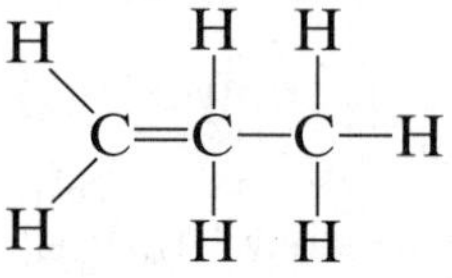
H   H H
  C=C-C-H
H   H H

____ 11. Which type of organic compound is represented by the structural formula shown below?

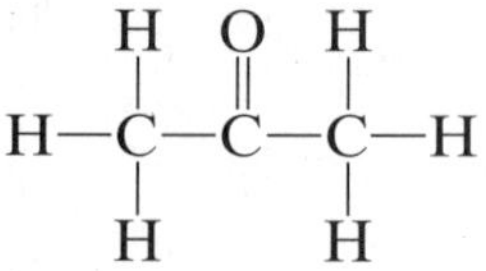

H   O   H
H-C-C-C-H
H   H

(1) aldehyde
(2) ketone
(3) alcohol
(4) organic acid

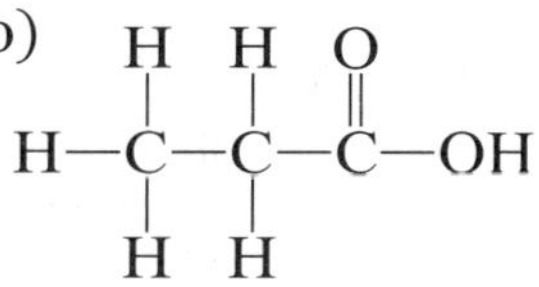

____ 12. Given the structural formulas of four organic compounds:

(a) 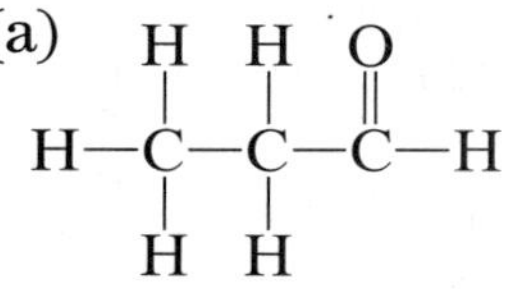
H H O
H-C-C-C-H
H H

(b)
H H O
H-C-C-C-OH
H H

(c)
H O H
H-C-C-C-H
H   H

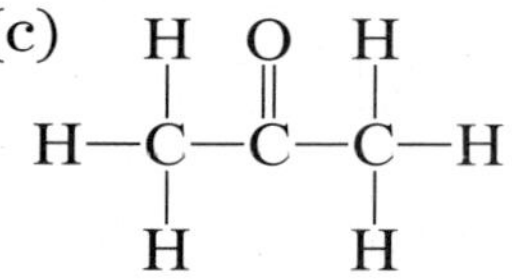

(d) 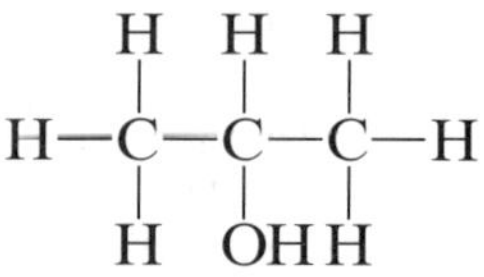
H H H
H-C-C-C-H
H OH H

Which pair below contains an alcohol and an aldehyde?
(1) a and b
(2) b and c
(3) b and d
(4) a and d

____ 13. Which of the following can have the greatest number of isomers?
(1) C_2H_6
(2) C_4H_{10}
(3) C_6H_6
(4) $C_{20}H_{42}$

____ 14. Which structural formula must be an alkene?
(1) C_4H_6
(2) C_4H_8
(3) C_4H_{10}
(4) C_6H_{14}

Unit XIII Organic Chemistry continued

_____ **15.** Given the following structural formula:

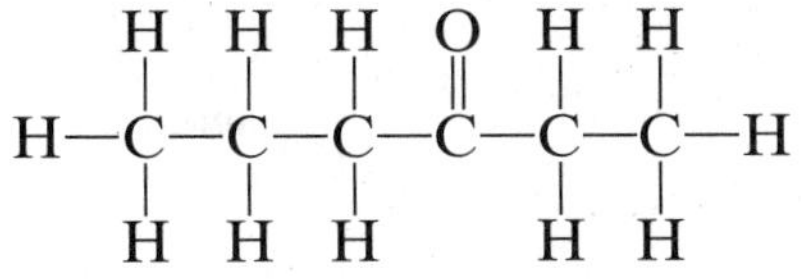

```
    H  H  H  O  H  H
    |  |  |  ||  |  |
H — C— C— C— C— C— C— H
    |  |  |     |  |
    H  H  H     H  H
```

Which is the correct name for the compound shown above?
(1) hexanal
(2) hexanol
(3) 3-hexanone
(4) 3-hexane

17. Draw the structural formula for butanoic acid.

18. Explain why isomers have different chemical and physical properties.

PART B-2

16. Given the structural formula for butane:

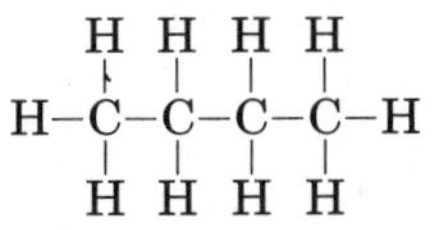

```
    H  H  H  H
    |  |  |  |
H — C— C— C— C— H
    |  |  |  |
    H  H  H  H
```

Draw the structural formula of an isomer of butane.

PART C

Trichlorofluoromethane, CCl_3F, was once used a refrigerant. Commonly known as *Freon-11*, it is no longer used because it was found to be destructive to the ozone layer. Trichlorofluoromethane was made by a substitution reaction.

19. Draw the structural formula for the hydrocarbon that was used as the basis for making *Freon-11*.

20. Draw the structural formula for *Freon-11*.

Name _____ Class _____ Date _____

UNIT XIV

Holt Chemistry: The Physical Setting

Oxidation and Reduction

You learned that an ionic bond forms when one or more electrons are transferred between atoms. An example is shown in the following equation.

$$2Na(s) + Cl_2(g) \longrightarrow 2NaCl(s)$$

Sodium chloride is a salt crystal made up of Na^+ cations and Cl^- anions. Therefore, even though the formula for sodium chloride is written as NaCl, you can also imagine it being written as Na^+Cl^-. You also learned that atoms are neutral. Therefore, sodium has no charge as a reactant, but it has a positive charge as a product. Similarly, chlorine has no charge as a reactant, but it has a negative charge as a product.

Sodium changed from Na^0 to Na^+ by losing one electron. The loss of one or more electrons from a substance is called **oxidation**. Chlorine changed from Cl^0 to Cl^- by gaining one electron. The gain of one or more electrons by a substance is called **reduction**. Oxidation and reduction occur together in a single reaction called an **oxidation-reduction reaction**, or redox reaction.

Redox Reactions

Although a redox reaction occurs as a single reaction, two **half-reactions** can be written to show how the electrons are transferred. A half-reaction represents the part of the reaction that involves only oxidation or reduction. For example, the half-reaction for the oxidation of sodium is written as follows.

$$Na(s) \longrightarrow Na^+(aq) + e^-$$

The half-reaction for the reduction of chlorine is written as follows.

$$Cl_2(g) + 2e^- \longrightarrow 2Cl^-(aq)$$

Notice that two chlorine atoms each gain an electron. Therefore, the half-reaction shown above involves the gain of two electrons. As a result the half-reaction for the oxidation of sodium must be rewritten to show the loss of two electrons.

$$2Na(s) \longrightarrow 2Na^+(aq) + 2e^-$$

When the two half-reactions are added, the net result should show the overall redox reaction.

What You'll Need to Learn

This topic is part of the Regents Curriculum for the Physical Setting Exam.
Standard 4, Performance Indicator 3.2: *Use atomic and molecular models to explain common chemical reactions.*
Standard 4, Performance Indicator 3.3: *Apply the principle of conservation of mass to chemical reactions.*

What Terms You'll Need to Know

anode
cathode
electrochemical cell
electrode
electrolysis
half-reaction
oxidation
oxidation number
oxidation-reduction reaction
reduction
voltaic cell

Where You Can Learn Even More:

Holt Chemistry: The Physical Setting Chapter 17: Oxidation, Reduction, and Electrochemistry

Unit XIV Oxidation and Reduction *continued*

Notes/Study Ideas/Answers

oxidation half-reaction: $2Na(s) \longrightarrow 2Na^+(aq) + 2e^-$

reduction half reaction: $Cl_2(g) + 2e^- \longrightarrow 2Cl^-(aq)$

redox reaction: $2Na(s) + Cl_2(g) \longrightarrow 2NaCl(s)$

Notice that the $2e^-$ on each side of the equation cancels one another. The half-reactions show that the number of electrons lost during oxidation equals the number of electrons gained during reduction.

> **Self-Check** Write the half reactions that show the loss of two electrons by Mg and the gain of two electrons by O.

You learned that the law of conservation of mass states that mass cannot be created or destroyed in ordinary physical and chemical changes. All the reactions shown above obey this law. Both Na and Cl are balanced with respect to mass. Notice also that redox half-reactions are balanced with respect to charge.

Take another look at the oxidation half-reaction.

$$2Na(s) \longrightarrow 2Na^+(aq) + 2e^-$$

Notice that the net charge on the left side of the equation is zero. The charges on the right side balance, leaving a net charge of zero. Now take another look at the reduction half-reaction.

$$Cl_2(g) + 2e^- \longrightarrow 2Cl^-(aq)$$

Notice that the net charge on the left side is 2–, and that the net charge on the right side is 2–.

> **Self-Check** Rewrite the following half-reaction so that it is balanced for mass and charge. $Br_2 + e^- \longrightarrow Br^-$

Looking at the half-reactions clearly reveals that both oxidation and reduction have taken place because the electrons are included in the equations. However, looking at an overall equation is not as revealing. For example, the following equation offers no information on whether it is a redox reaction.

$$C(s) + O_2(g) \longrightarrow CO_2(g)$$

The reaction shown above is a clearly an example of a combustion reaction in which carbon burns in oxygen to produce carbon dioxide. However, is this reaction also a redox reaction? The only way to determine the answer is to assign **oxidation numbers** to each atom. An oxidation number represents the number of electrons that must be added or removed to an atom in a bonded state to make it neutral.

Unit XIV Oxidation and Reduction *continued*

Let's begin by assigning oxidation numbers to the atoms in the reactants. As an atom, C is neutral. No electrons must be added or subtracted to make it neutral. Therefore, C is assigned an oxidation number of 0. All compounds are neutral. Therefore, each O atom in O_2 has an oxidation number of 0.

Next, let's assign oxidation numbers to the atoms in the product. Each O atom in CO_2 has gained two electrons. To make each O atom neutral, two electrons would have to be removed. As a result each O atom is assigned an oxidation number of 2–.

The bond in CO_2 is a polar covalent bond. The shared electrons spend more time around each O atom. You can think of the C atom as losing its four electrons part of the time. To become neutral, the C atom must gain four electrons. Therefore, the oxidation number of C in CO_2 is 4+.

Self-Check How have the oxidation numbers changed in the following reaction? $C(s) + O_2(g) \longrightarrow CO_2(g)$

A simpler way to determine the oxidation number of C in CO_2 is to consider the total charge contributed by the two O atoms. If each O atom has an oxidation number of 2–, then two O atoms must have a total oxidation number charge of 4–. All compounds are neutral. Therefore, the oxidation number of the C atom in CO_2 must be 4+ to balance the 4– from the two O atoms.

The simplest way to determine the oxidation number of an atom is to check a periodic table that lists oxidation states. For example, if you check the periodic table provided with a Regents exam, you will see that the oxidation state for O is –2, which is another way of writing 2–.

Some elements have more than one oxidation state listed. For example, Cl has three oxidation states listed. The first one listed is the most common oxidation state or number.

Self-Check What are the oxidation numbers for F, Li, and Ca?

There are rules that apply to assigning oxidation numbers. One rule states that the oxidation number of an atom of any uncombined element in atomic or molecular form is zero. Therefore, the oxidation number for each atom in Ne, Ar, H_2, Cl_2, O_2, and S_8 is zero.

Another rule states that the oxidation number of a monatomic ion is equal to the charge on the ion. Therefore, the oxidation number of K^+ is 1+, and the oxidation number of Al^{3+} is 3+.

Unit XIV Oxidation and Reduction *continued*

Self-Check What are the oxidation numbers for F_2, He, and Fe^{3+}?

Once all oxidation numbers have been assigned, you can then tell if a redox reaction has occurred. For example, consider the following reaction.

$$H_2(g) + CuO(s) \rightarrow Cu(s) + H_2O(l)$$

Based on the rules you just read, both H_2 and Cu are assigned oxidation numbers of zero.

The O atom in CuO has an oxidation number of 2–. Therefore, the Cu atom must have an oxidation number of 2+.

The O atom in H_2O has an oxidation number of 2–. Therefore, the oxidation number for *both* H atoms must be 2+. Each H atom then has an oxidation number of 1+.

The oxidation numbers are usually written above each atom in the reaction.

$$\overset{0}{H_2}(g) + \overset{2+\ 2-}{CuO}(s) \rightarrow \overset{0}{Cu}(s) + \overset{1+\ 2-}{H_2O}(l)$$

Notice that H has changed its oxidation number from 0 to 1+. Also notice that Cu has changed its oxidation number from 2+ to 0. These changes in oxidation numbers indicate that a redox reaction has occurred.

REVIEW YOUR UNDERSTANDING

_____ **1.** Oxidation is defined as the
  (1) gain of electrons.
  (2) loss of electrons.
  (3) transfer of electrons.
  (4) loss of charge.

_____ **2.** Which half-reaction represents reduction?
  (1) $K \rightarrow K^+ + e^-$
  (2) $2H_2 + O_2 \rightarrow 2H_2O$
  (3) $S + 2e^- \rightarrow S^{2-}$
  (4) $2K + S \rightarrow K_2S$

_____ **3.** Which substance has an oxidation number of zero?
  (1) Ba^{2+}
  (2) PO_4^{2-}
  (3) a magnesium cation
  (4) I_2

Unit XIV Oxidation and Reduction continued

_____ 4. What is the oxidation number of Na in Na_2O?
   (1) 1+
   (2) 2+
   (3) 1–
   (4) 2–

_____ 5. Which half-reaction is correctly balanced for both mass and charge?
   (1) $2KNO_3 \longrightarrow 2KNO_2 + O_2$
   (2) $Fe(s) \longrightarrow Fe^{3+}(aq) + 3e^-$
   (3) $O_2(g) + e^- \longrightarrow O^{2-}(aq)$
   (4) $Cu^+(aq) + 2e^- \longrightarrow Cu(s)$

Notes/Study Ideas/Answers

Electrochemical Cells

Flashlights work because of redox reactions that occur in batteries. A battery is an **electrochemical cell**, which is a device that consists of two **electrodes** separated by an electrolyte. You learned that an electrolyte is a substance that conducts electricity. An electrode is a conductor that connects with a nonmetallic part of a circuit, such as an electrolyte. These definitions involving an electrochemical cell will make better sense by taking a look at what happens inside a battery.

The redox reactions that occur inside a battery are rather involved. For example, a half-reaction that occurs in a dry-cell battery used in a flashlight is written as follows.

$$2MnO_2(s) + 2NH_4^+(aq) + 2e^- \longrightarrow Mn_2O_3(s) + 2NH_3(aq) + H_2O(l)$$

For this reason, we will look at a simpler redox reaction that can serve as a model of what happens inside a battery.

Self-Check Does the half-reaction shown above represent oxidation or reduction?

Unit XIV Oxidation and Reduction continued

As an electrochemical cell, a battery changes chemical energy into electrical energy. The following illustration can serve as a model of how a battery operates as an electrochemical cell.

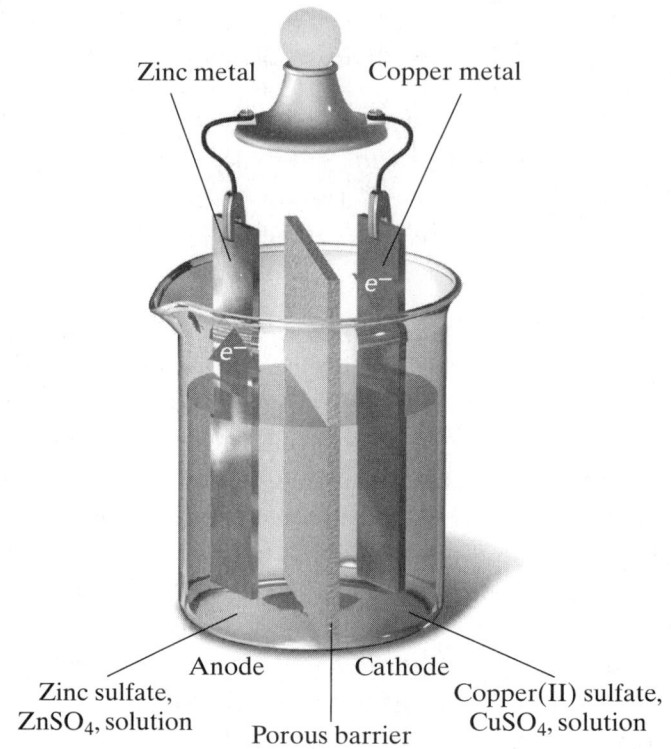

Notice that the electrochemical cell includes two metal strips, which serve as the electrodes. The zinc metal electrode is immersed in a solution of zinc sulfate, an electrolyte. The copper metal electrode is immersed in a solution of copper(II) sulfate, another electrolyte. Both electrodes are connected to a light bulb by wires.

A porous barrier separates the two electrodes and keeps the electrolytic solutions from mixing. The ions in solution, however, can pass through the porous barrier. For example, Zn^{2+} ions can move from left to right, and Cu^{2+} ions can move from right to left.

The atoms in the zinc electrode lose electrons. The half-reaction is written as follows.

$$Zn(s) \longrightarrow Zn^{2+}(aq) + 2e^-$$

Therefore, Zn(s) undergoes oxidation to form $Zn^{2+}(aq)$. The electrode at which oxidation occurs is called the **anode**. As zinc atoms lose electrons to form ions at the anode, the zinc strip slowly dissolves.

Unit XIV Oxidation and Reduction *continued*

The copper ions in solution near the copper electrode gain the electrons lost by the zinc atoms. The half-reaction is written as follows.

$$Cu^{2+}(aq) + 2e^- \rightarrow Cu(s)$$

Therefore, $Cu^{2+}(aq)$ undergoes reduction to form $Cu(s)$. The electrode at which reduction occurs is called the **cathode**. As $Cu(s)$ forms at the cathode, the copper atoms are deposited on the metal strip.

The electrons lost by $Zn(s)$ and gained by $Cu^{2+}(aq)$ pass through the wire. The flow of electrons creates the electricity that lights up the bulb.

Self-Check Eventually, the light bulb will no longer be lit. Explain why.

In the redox reaction you just examined, Zn atoms in the zinc electrode lose electrons to the Cu^{2+} ions in the copper(II) sulfate solution. But why don't Cu atoms in the copper electrode lose electrons to the Zn^{2+} ions in the zinc sulfate solution? The answer can be found in *Reference Table J*, which contains a list of elements that arranged in an activity series.

An activity series shows the relative activity of the substances listed on the table. Notice in *Reference Table J* that the metals are listed in the left column, while the nonmetals are listed in the right column. The metal with the highest activity, Li, is listed at the top. As you proceed down the column, the activity of the metals decreases. Gold (Au) has the least activity.

The activity of the metals refers to their tendency to undergo oxidation or give up electrons. Notice that Zn is listed above Cu in *Reference Table J*. Therefore, Zn will undergo oxidation more readily than Cu. As a result, Zn atoms give up their electrons to Cu^{2+} ions rather than Cu atoms giving up their electrons to Zn^{2+} ions.

Self-Check Which undergoes oxidation more readily, aluminum or silver?

In effect, the redox reaction you just examined is a single-replacement reaction.

$$Zn(s) + CuSO_4(aq) \rightarrow Cu(s) + ZnSO_4(aq)$$

An activity series then is useful for predicting if a single-replacement reaction will occur spontaneously. For example, consider what happens when $Mg(s)$ and $Pb(NO_3)_2(aq)$ are mixed. Will Mg replace Pb? If so, then the following reaction should be spontaneous.

Notes/Study Ideas/Answers

$$Mg(s) + Pb(NO_3)_2(aq) \longrightarrow Pb(s) + Mg(NO_3)_2(aq)$$

In the reaction above, Mg undergoes the following change in oxidation numbers.

$$\overset{0}{Mg}(s) + Pb(NO_3)_2(aq) \longrightarrow Pb(s) + \overset{2+}{Mg}(NO_3)_2(aq)$$

Therefore, Mg undergoes oxidation by losing two electrons. If the reaction above occurs spontaneously, then Mg must be more likely to undergo oxidation than Pb. If you check *Reference Table J*, you will see that Mg is listed above Pb. As a result, the single-replacement reaction shown above does occur spontaneously.

You learned that an electrochemical cell can spontaneously change chemical energy into electrical energy. This type of electrochemical cell is called a **voltaic cell**.

Some electrochemical cells use electrical energy to bring about a chemical change. These electrochemical cells are called **electrolytic cells**. For example, the illustration shown on this page illustrates how an electrolytic cell can be used to break down water. Notice that an electrolytic cell requires electrical energy that is supplied by an external source.

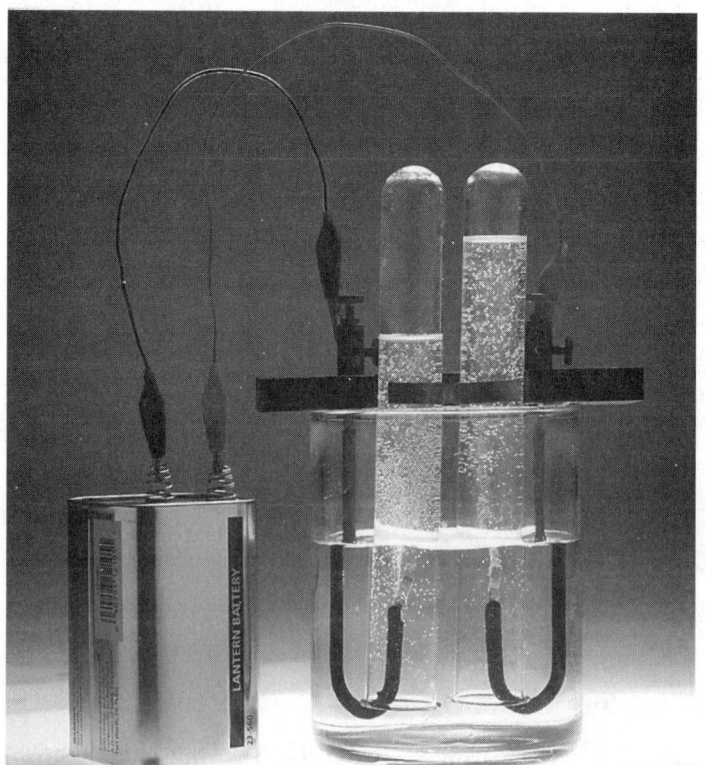

Unit XIV Oxidation and Reduction *continued*

Notice that the electrical energy used by the electrolytic cell to break down water is supplied by a battery (voltaic cell). The process carried out by an electrolytic cell is called **electrolysis**. Electrolysis is the process by which an electric current is used to produce a chemical change, such as the decomposition of water.

Electrolysis is also used to plate various metal objects with other metals to give them the appearance of silver or gold. For example, a tin bracelet can be made to look like pure silver by simply coating it with a thin layer of silver. The following illustration shows how electrolysis is carried out to plate a bracelet with silver.

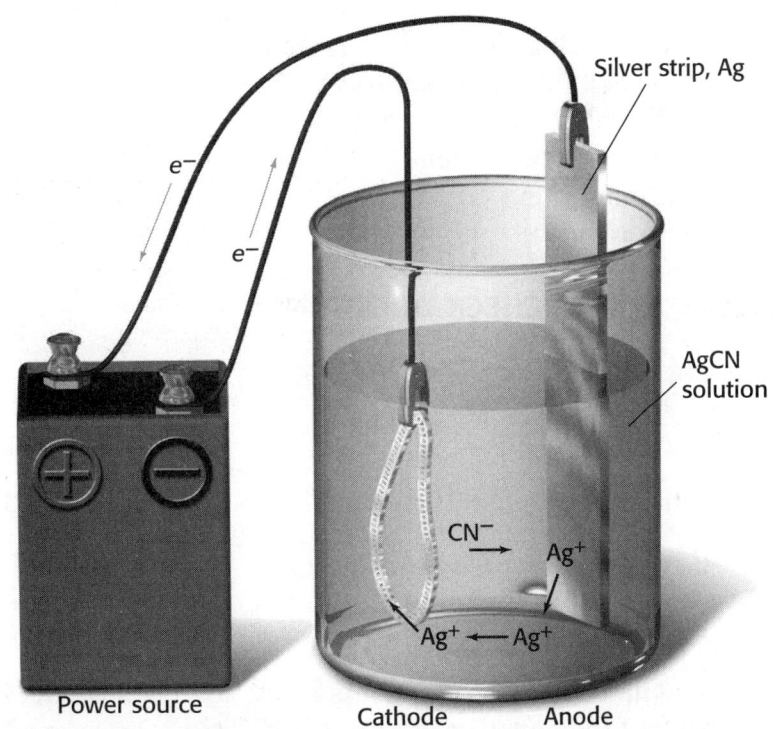

REVIEW YOUR UNDERSTANDING

_____ 6. In a voltaic cell, oxidation occurs at
   (1) the anode.
   (2) the cathode.
   (3) either the anode or the cathode.
   (4) both the anode and cathode simultaneously.

_____ 7. Which half-reaction is correctly written to show what happens at the cathode in a voltaic cell?
   (1) $Zn(s) \rightarrow Zn^{2+}(aq) + 2e^-$
   (2) $Cu^{2+}(aq) + e^- \rightarrow Cu(s)$
   (3) $Fe^{3+}(aq) + 3e^- \rightarrow Fe^{2+}(aq)$
   (4) $2H^+(aq) + 2e^- \rightarrow H_2(g)$

Unit XIV Oxidation and Reduction *continued*

_____ **8.** Which metal will replace zinc in a single-replacement reaction?
(1) copper
(2) nickel
(3) magnesium
(4) gold

_____ **9.** In an electrolytic cell,
(1) chemical energy is changed into electrical energy.
(2) reduction occurs when a metal loses electrons.
(3) electrical energy is used to bring about a chemical change.
(4) the redox reaction occurs spontaneously.

_____ **10.** When a battery is being recharged, the electrochemical cell is acting as a(n)
(1) voltaic cell.
(2) electrolytic cell.
(3) cathode.
(4) anode.

ANSWERS TO SELF-CHECK QUESTIONS

- $Mg \rightarrow Mg^{2+} + 2e^-$; $O_2 + 2e^- \rightarrow 2O^{2-}$

- $Br_2 + 2e^- \rightarrow 2Br^-$

- C has gone from 0 to 4+, and O has gone from 0 to 2−.

- F is −1. Li is +1. Ca is +2.

- F_2 is 0. He is 0. Fe^{3+} is 3+.

- The half-reaction represents reduction because the left side of the equation shows that electrons are gained.

- The redox reaction will cease to operate, most likely because the zinc metal is completely dissolved when all the zinc atoms have been oxidized.

- aluminum

- The reaction will not occur spontaneously because Al is listed below K in *Reference Table J*.

UNIT XIV

Questions for Regents Practice

Holt Chemistry: The Physical Setting

PART A

_____ 1. Which particles are gained and lost during a redox reaction?
(1) protons
(2) neutrons
(3) electrons
(4) both electrons and protons so that neutrality is maintained.

_____ 2. What is the oxidation number of Cr in $K_2Cr_2O_7$?
(1) −6
(2) +1
(3) +6
(4) +12

_____ 3. Given the reaction:

$Mg(s) + 2H^+(aq) + 2Cl^-(aq) \rightarrow Mg^{2+}(aq) + 2Cl^-(aq) + H_2(g)$

Which substance undergoes reduction?
(1) $Mg(s)$
(2) $H^+(aq)$
(3) $Cl^-(aq)$
(4) $H_2(g)$

_____ 4. In which compound does chlorine have the highest oxidation number?
(1) KClO
(2) $KClO_2$
(3) $KClO_3$
(4) $KClO_4$

_____ 5. When a neutral atom undergoes reduction, its oxidation number
(1) increases.
(2) decreases.
(3) stays the same.
(4) becomes zero.

_____ 6. In any redox reaction, the substance that undergoes oxidation will
(1) lose electrons and have an increase in oxidation number.
(2) lose electrons and have a decrease in oxidation number.
(3) gain electrons and have an increase in oxidation number.
(4) gain electrons and have a decrease in oxidation number.

Questions 7–9 apply to the following reaction.

$Zn(s) + 2HCl(aq) \rightarrow ZnCl_2(aq) + H_2(g)$

_____ 7. Which statement about this redox reaction is correct?
(1) Mass, but not charge, is conserved.
(2) Charge, but not mass, is conserved.
(3) Atoms of $Zn(s)$ undergo oxidation.
(4) Atoms of $Zn(s)$ gain electrons.

_____ 8. What substance undergoes a decrease in oxidation number?
(1) $Zn(s)$
(2) H in $HCl(aq)$
(3) $H_2(g)$
(4) Zn in $ZnCl_2(aq)$

_____ 9. What is the change in oxidation number for Zn?
(1) 0 to +2
(2) +2 to 0
(3) −2 to +2
(4) +4 to +2

Copyright © by Holt, Rinehart and Winston. All rights reserved.

Oxidation and Reduction

Unit XIV Oxidation and Reduction continued

PART B-1

Questions 10–11 apply to the following reaction that shows how pure aluminum is obtained by electrolysis.

$$2Al_2O_3(l) + 3C(s) \longrightarrow 4Al(l) + 3CO_2(g)$$

_____ 10. Which is a half-reaction that occurs in the reaction shown above?
  (1) $C + 4e^- \longrightarrow C^{4+}$
  (2) $C \longrightarrow C^{2+} + 2e^-$
  (3) $Al^{2+} + 2e^- \longrightarrow Al$
  (4) $Al^{3+} + 3e^- \longrightarrow Al$

_____ 11. The C(s) in the redox reaction above undergoes
  (1) reduction by gaining two electrons.
  (2) reduction by gaining four electrons.
  (3) oxidation by losing two electrons.
  (4) oxidation by losing four electrons.

Questions 12-14 apply to the following illustration.

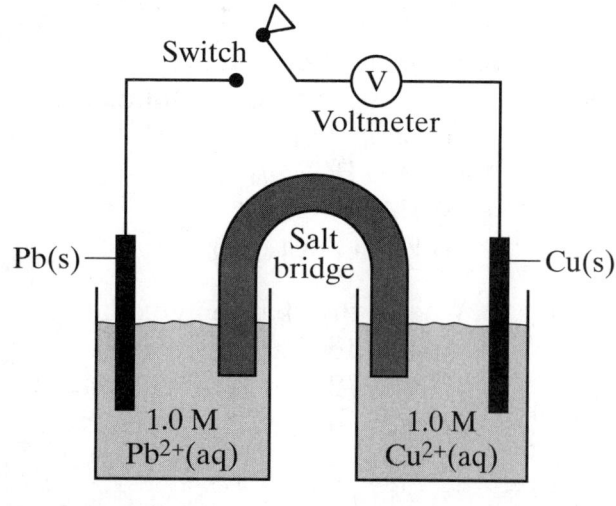

Pb(s) + Cu²⁺(aq) ⟶ Pb²⁺(aq) + Cu(s)

_____ 12. In what direction will electrons flow when the switch is closed?
  (1) from Cu(s) to Pb(s)
  (2) from Pb(s) to Cu(s)
  (3) from Pb(s) to Pb²⁺(aq)
  (4) from Cu(s) to Cu²⁺(aq)

_____ 13. What is the purpose of the salt bridge?
  (1) It allows electrons to flow between the electrodes.
  (2) It is the cathode where reduction occurs.
  (3) It is a path for the flow of cations and anions.
  (4) It prevents the Pb from oxidizing.

_____ 14. What will happen over time?
  (1) The lead strip will slowly dissolve.
  (2) The lead strip will slowly accumulate more lead atoms.
  (3) The copper strip will slowly dissolve.
  (4) All the Pb²⁺ ions will be reduced to form Pb atoms.

PART B-2

15. Explain how a voltaic cell differs from an electrolytic cell.

_____
_____
_____

16. Is the reaction below a redox reaction? Explain your answer.

$$2HCl + Mg(OH)_2 \longrightarrow MgCl_2 + 2H_2O$$

_____
_____
_____
_____

Unit XIV Oxidation and Reduction *continued*

17. Nitrogen monoxide, NO(*g*), reacts with phosphorus, $P_4(s)$, to produce nitrogen, $N_2(g)$, and diphosphorus pentoxide, P_2O_5. Write the balanced equation for this reaction. Identify the atoms that have been oxidized and reduced. Also show the changes in their oxidation numbers.

_____

_____

_____

PART C

Fuel cells are being investigated as a source of unlimited energy. Unlike a battery, a fuel cell can work, in theory, forever, changing chemical energy into electrical energy. One type of fuel cell uses methanol, CH_3OH, as a fuel. The fuel reacts with oxygen gas, O_2, to produce CO_2 and H_2O.

18. Write the balanced equation for the reaction that occurs in this fuel cell.

19. Write the oxidation half-reaction. Be sure to balance both **mass and charge**.

20. Write the reduction half-reaction. Be sure to balance both **mass and charge**.

UNIT XV
Nuclear Chemistry

Holt Chemistry: The Physical Setting

The chemical reactions you have examined so far involve electrons. During these chemical reactions, atoms are rearranged when bonds are broken and new ones are formed. You learned that chemical reactions obey two laws: the law of conservation of mass and the law of conservation of energy. In addition, you learned in your study of redox reactions that charge is also conserved during a reaction

However, not all reactions involve electrons. In addition, not all reactions result in products that have a mass equal to that of the reactants. Rather, some reactions involve the nuclei of atoms and result in mass being converted into energy. These types of reactions are called nuclear reactions. Nuclear reactions involve more than just a rearrangement of atoms. Nuclear reactions involve the creation of new substances. These nuclear reactions are called **transmutations** in which one element changes into another element.

Transmutations

Some transmutations occur naturally. These transmutations happen because some nuclei are unstable. The ratio of the neutrons and protons in an atom determines the stability of its nucleus. Unstable nuclei spontaneously break down or decay in a process called **radioactivity**. Radioactivity involves the emission of one or more particles or energy in the form of electromagnetic radiation.

Consider what happens to a carbon-14 atom, whose nucleus is unstable. Carbon-14 undergoes a transmutation as shown by the following equation.

$$^{14}_{6}C \longrightarrow {}^{14}_{7}N + {}^{0}_{-1}e$$

The equation above represents a transmutation in which a carbon-14 atom changes into a nitrogen-14 atom. As a result of this transmutation, a neutron has changed into a proton. Whenever this happens, a high-energy electron, called a **beta particle**, is emitted. The beta particle has the notation $^{0}_{-1}e$.

If you again examine the equation above, you will notice that the numbers on each side are balanced. The superscripts, which represent mass, add up to 14 on each side. The subscripts, which represent charge, add up to 6 on each side.

What You'll Need to Learn

This topic is part of the Regents Curriculum for the Physical Setting Exam.
Standard 4, Performance Indicator 4.4: *Explain the benefits and risks of radioactivity.*
Standard 4, Performance Indicator 5.3: *Compare energy relationships within an atom's nucleus to those outside the nucleus.*

What Terms You'll Need to Know

alpha particle
beta particle
gamma radiation
half-life
nuclear fission
nuclear fusion
radioactivity

Where You Can Learn Even More:

Holt Chemistry: The Physical Setting
Chapter 18: Nuclear Chemistry

Self-Check Write the notation for a beta particle.

Unit XV Nuclear Chemistry continued

A balanced nuclear equation is useful if either a reactant or product is unknown. For example, consider the following equation.

$$^{38}_{19}K \longrightarrow ? + ^{0}_{-1}e$$

Notice that an atom of potassium-38 spontaneously decays and releases a beta particle. However, the identity of the other product is unknown. For the subscripts to be balanced, the subscript on the unknown product must be 20.

You learned that the subscript in an atomic symbol represents the atomic number. Checking a periodic table reveals that the element with atomic number 20 is calcium, Ca. Therefore, in the equation shown above, potassium-38 decays to form a calcium atom.

For the masses to be balanced, the superscript on Ca must be 38. Therefore, the completed equation is written as follows.

$$^{38}_{19}K \longrightarrow ^{38}_{20}Ca + ^{0}_{-1}e$$

Self-Check Complete the following nuclear equation.
$$? \longrightarrow ^{234}_{91}Pa + ^{0}_{-1}e$$

Some unstable nuclei release **alpha particles** when they decay. An alpha particle is a positively-charged particle that consists of two protons and two neutrons. This same arrangement of subatomic particles is found in a helium nucleus. Therefore, the notation used for an alpha particle is written as $^{4}_{2}He$. *Reference Table O* lists the notations and symbols used for the various particles involved in nuclear chemistry.

Although He is used as part of the notation, keep in mind that $^{4}_{2}He$ does not represent a helium atom but rather an alpha particle (helium nucleus). An example of an alpha decay is shown in the following equation.

$$^{238}_{92}U \longrightarrow ^{234}_{90}Th + ^{4}_{2}He$$

The equation above represents a transmutation in which an atom or uranium-238 emits an alpha particle and forms an atom of thorium-234.

Self-Check Complete the following nuclear equation.
$$? \longrightarrow ^{208}_{82}Pb + ^{4}_{2}He$$

As you can see from *Reference Table O*, there are other particles involved in nuclear chemistry. By knowing that a nuclear equation must be balanced for mass and charge, you can always determine the identity of an unknown substance. For example, consider the following spontaneous decay reaction.

Unit XV Nuclear Chemistry continued

$$^{49}_{24}Cr \longrightarrow \, ^{49}_{23}V + ?$$

To balance the charges, the unknown must have a +1 as a subscript. To balance the masses, the unknown must have a superscript of 0. If you check *Reference Table O*, you will see that the unknown substance must be a particle called a positron, whose notation is written as $^{0}_{+1}e$.

Self-Check Complete the following nuclear equation. $? \longrightarrow \, ^{15}_{7}N + \, ^{0}_{+1}e$

Notice in *Reference Table O* that nuclear reactions also release **gamma radiation**. Gamma radiation consists of high-energy waves that have no mass or charge. Gamma rays can only be stopped by several centimeters of lead. In contrast, the skin or one sheet of paper can stop an alpha particle, while a few sheets of aluminum foil can stop a beta particle.

The transmutations that you have studied so far occur spontaneously in nature. However, there are transmutations that scientists carry out in laboratories with the help of powerful devices called particle accelerators. These devices generate tremendous quantities of energy to accelerate particles, such as alpha particles, to 99.9999% of the speed of light.

Traveling at such high speed, the particles are then bombarded into targets, which may consist of atoms. An example is shown in the following equation, which shows what can happen when a neon atom is bombarded with a high-speed alpha particle.

$$^{21}_{10}Ne + \, ^{4}_{2}He \longrightarrow \, ^{24}_{12}Mg + \, ^{1}_{0}n$$

Notice that the neon atom has been changed into a magnesium atom. This is an example of an artificial transmutation. The other product is a neutron, $^{1}_{0}n$.

REVIEW YOUR UNDERSTANDING

_____ 1. Which particle consists of two protons and two neutrons?
   (1) alpha particle
   (2) beta particle
   (3) positron
   (4) gamma radiation

_____ 2. Gamma rays
   (1) have the same energy as alpha particles.
   (2) are visible light.
   (3) have no mass and no charge.
   (4) can be stopped by several pieces of aluminum foil.

Notes/Study Ideas/Answers

Unit XV Nuclear Chemistry continued

_____ 3. A nuclear equation must be balanced for
(1) mass, only.
(2) charge, only.
(3) mass and charge, both.
(4) the number of particles involved in the reaction.

_____ 4. Identify the unknown in the following equation.

$$^{239}_{93}\text{Np} \longrightarrow ^{0}_{-1}\text{e} + ?$$

(1) $^{239}_{92}\text{U}$
(2) $^{238}_{92}\text{U}$
(3) $^{93}_{239}\text{Np}$
(4) $^{239}_{94}\text{Pu}$

_____ 5. When a radioactive atom emits a beta particle, its
(1) atomic number decreases by one.
(2) atomic number increases by one.
(3) atomic number remains the same.
(4) mass number decreases by one.

Fission and Fusion

You have studied nuclear reactions involving both natural and artificial transmutations. Another type of nuclear reaction is called **nuclear fission**. Nuclear fission is the splitting of the nucleus of a large atom into two or more fragments. Some unstable nuclei undergo spontaneous fission. However, most fission reactions happen when a nucleus is bombarded with neutrons.

Consider the following fission reaction.

$$^{235}_{92}\text{U} + ^{1}_{0}\text{n} \longrightarrow ^{93}_{36}\text{Kr} + ^{140}_{56}\text{Ba} + 3^{1}_{0}\text{n}$$

In the reaction above, an atom of uranium-235 is bombarded with a neutron. The products include krypton-93, barium-140, and three neutrons. The three neutrons produced in the reaction above can bombard additional unstable nuclei to continue the fission process. This is known as a chain reaction. One characteristic of a chain reaction is that the particle that starts the reaction, in this case a neutron, is also a product of the reaction.

Self-Check A neutron produced in the reaction above can bombard an atom of uranium-235 that then undergoes fission. The products include Br-87, La-146, and two neutrons. Write the balanced equation for this fission reaction.

Unit XV Nuclear Chemistry continued

The total mass of the products in the fission reaction does not equal the total mass of the reactants. This would seem to violate the law of conservation of mass. However, mass has not been destroyed. Rather, mass has been converted into energy.

Albert Einstein described the equivalence of mass and energy in his famous equation $E = mc^2$, where E represents energy, m represents mass, and c is the speed of light. This equation shows that even a small amount of mass is converted to a significant quantity of energy.

A fission reaction involves many more than just one neutron and one atom of uranium-235. As a result, a tremendous quantity of energy is released by a fission reaction. Energy released during nuclear reactions is much greater than the energy released during chemical reactions.

Self-Check Explain why the difference between the total mass of the products and the total mass of the reactants is known as the mass defect or mass loss.

The conversion of mass into energy also occurs during a **nuclear fusion** reaction. Nuclear fusion is the combination of the nuclei of small atoms to form a larger nucleus, a process that releases a tremendous quantity of energy.

Nuclear fusion is the process by which the stars, including our sun, generate energy. In the sun, four hydrogen nuclei fuse to form a single helium nucleus, as shown in the following equation.

$$4\,^{1}_{1}H \longrightarrow \,^{4}_{2}He + 2\,^{0}_{+1}e$$

Extremely high temperatures are required to bring about a fusion reaction. For example, the temperature of the sun's core, where some of the fusion reactions occur, is about 15,000,000°C (1.5×10^7°C).

Scientists are currently investigating fusion reactions as a source of energy. So far, researchers use just as much energy to start the fusion reaction as is released by the reaction. As a result, fusion is not yet a practical source of energy.

REVIEW YOUR UNDERSTANDING

_____ 6. The combining of several smaller nuclei to form a larger nucleus is called
   (1) nuclear fission.
   (2) nuclear fusion.
   (3) alpha emission.
   (4) beta emission.

Notes/Study Ideas/Answers

Unit XV Nuclear Chemistry continued

_____ 7. Both nuclear fission and nuclear fusion
   (1) emit alpha particles.
   (2) occur in the stars.
   (3) release tremendous quantities of energy.
   (4) combine hydrogen nuclei to form a helium nucleus.

_____ 8. A nucleus of Cm-246 can be combined with a nucleus of C-12. The nucleus that is produced in this reaction is
   (1) Fm-254.
   (2) Es-254.
   (3) Md-256.
   (4) No-258.

_____ 9. Einstein's equation, $E = mc^2$, shows that
   (1) mass and energy are equivalent.
   (2) mass is always conserved in any reaction.
   (3) energy cannot be converted to mass.
   (4) nuclear reactions release less energy than chemical reactions.

Uses of Nuclear Chemistry

Nuclear chemistry can be used for various practical purposes. One of its uses is dating the age of specimens. This dating process is possible because radioactive substances undergo nuclear decay at a constant rate. This rate of nuclear decay is measured in terms of its **half-life**. A half-life is the time required for half of a sample of radioactive substance to decay by natural processes.

The half-life of a radioactive substance is constant and is not influenced by external conditions, such as temperature and pressure. Each half-life varies, depending on the radioactive substance. For example, iodine-131 has a half-life of just over eight days. In contrast, uranium-235 has a half-life of 704,000,000 years (7.04×10^8 years).

Reference Table N lists the half-lifes for selected radioisotopes, or atoms that are radioactive. Notice that this table also lists the decay mode for each radioisotope. For example, nitrogen-16, decays by emitting a beta particle and has a half-life of 7.2 seconds.

Self-Check What is the half-life of cesium-137? Write a balanced equation to show how cesium-137 decays.

Unit XV Nuclear Chemistry continued

One of the most useful radioisotopes for dating specimens is carbon-14. The half-life of carbon-14 is 5730 years. After that interval, only half of the original carbon-14 will remain. In another 5730 years, half of the remaining carbon-14 atoms will have decayed and leave one-fourth of the original amount.

Scientists measure the ratio of carbon-14 to carbon-12 in a specimen they want to date. By comparing this ratio to the ratio in a sample of similar material whose age is known, scientists can determine the age of the specimen. For example, suppose an artifact is found to halve one-eighth the ratio of carbon-14 to carbon-12 compared to the ratio in a similar object today.

For the artifact to have one-eighth the ratio of carbon-14 to carbon-12, three half-lifes must have elapsed.

$$\frac{1}{2} \times \frac{1}{2} \times \frac{1}{2} = \frac{1}{8}$$

To determine the age of the artifact, multiply the half-life of carbon-14 three times for the three half-lifes that have elapsed.

$$3 \times 5730 \text{ years} = 17{,}190 \text{ years}$$

Self-Check If you start with 16 milligrams of strontium-90, how much of the original radioisotope will remain after 112.4 years?

You can determine the age of a specimen, how much remains of a radioisotope, the amount that was present in the original sample, or even the half-life of a radioisotope if you are given certain information.

For example, suppose you are told that a sample contains 2 mg of radium-226 and was found to be 9600 years old. How much radium-226 was present in the original sample?

Reference Table N lists the half-life of radium-226 as 1600 years.

$$\frac{9600 \text{ years}}{1600 \text{ years}} = 6 \text{ half-lifes have elapsed}$$

You must double the amount of radium-226 to determine how much existed for each half-life that has elapsed.

2 mg × 2 mg × 2 mg × 2 mg × 2 mg × 2 mg × 2 mg = 128 mg existed in the original sample 9600 years ago

Self-Check Phosphorus-32 is used to treat certain forms of leukemia. Scientists determined that a specimen once contained 100.0 mg of phosphorus-32 and was dated as being 57.2 days old. Today, the specimen contains only 6.25 mg of this radioisotope. What is the half-life of phosphorus-32?

Notes/Study Ideas/Answers

Unit XV Nuclear Chemistry continued

In addition to dating specimens, nuclear chemistry has also been put to other uses. Fission reactions generate the electrical energy in nuclear reactors. Uranium-235 and plutonium-239 are the main radioisotopes used in the reactors. The chain reactions generate a tremendous quantity of energy as heat that is used to turn water into steam. The steam then drives a turbine to produce electrical energy.

Control rods are used to adjust the rate of the chain reactions so that they do not get out of control. In 1986, technicians briefly removed most of the reactor's control rods in a nuclear reactor in Chernobyl in the Ukraine during a safety test. Such accidents are rare, but their impact can be quite serious.

Another concern about nuclear reactors concerns the fuel rods that are loaded with the radioisotopes. At some point, the radioisotopes are no longer able to sustain a chain reaction and must be removed. The rods are known as "spent fuel rods" and are considered nuclear waste.

At present, there is no central depository for the safe storage of spent fuel rods from all the reactors in the country. As a result, nuclear wastes are often stored in "spent-rod pools" where they are covered with at least six meters of water. This amount of water prevents radiation from the spent fuel rods from harming people.

Radiation can affect people in several ways, depending on how much exposure a person has experienced. The unit for exposure is the *rem*, which expresses the biological effect of an absorbed dose of radiation in humans. The following table summarizes the effect of whole-body exposure to a single dose of radiation.

Table 3 Effect of Whole-Body Exposure to a Single Dose of Radiation

Dose (rem)	Probable effect
0–25	no observable effect
25–50	slight decrease in white blood cell count
50–100	marked decrease in white blood cell count
100–200	nausea, loss of hair
200–500	ulcers, internal bleeding
> 500	death

Self-Check How might an exposure of a single dose of 200 rem affect a person?

Unit XV Nuclear Chemistry continued

People who work in areas where radioisotopes are used wear a badge that monitors their exposure. Healthcare professionals are advised to limit their exposure to 5 rem per year. This level of exposure is 1000 times higher than the recommended exposure for most people, including you.

The radioisotopes that healthcare workers are exposed to are used for the diagnosis and treatment of various diseases. For example, thallium-201 is used to produce an image of a person's heart to check its condition. Technetium-99 is used for bone scans. Cobalt-60 is used to treat cancerous tumors.

The benefits of nuclear chemistry can also be found in most homes. Smoke detectors depend on a nuclear reaction to sound an alarm when a fire starts. Many smoke detectors contain a small amount of americium-241, which decays by emitting an alpha particle.

Self-Check Write the nuclear equation that shows the decay of americium-214.

The alpha particles emitted by americium-241 ionize the air inside the smoke detector. This changes the molecules into ions, which conduct an electric current. Smoke particles reduce this current when they mix with the ionized air. In response, the smoke detector sets off an alarm. The americium-241 is harmless because the alpha particles cannot pass through the plastic covering. Even if they do, the alpha particles can travel only a short distance.

REVIEW YOUR UNDERSTANDING

_____ 10. The half-life of thorium-234 is 24 days. If you start with a 42.0 g sample of this radioisotope, how much will remain after 96 days?
  (1) 2.63 g
  (2) 5.25 g
  (3) 10.5 g
  (4) 21.0 g

_____ 11. Which radioisotope is used in nuclear reactors?
  (1) polonium-212
  (2) thorium-234
  (3) uranium-235
  (4) americium-241

Unit XV Nuclear Chemistry continued

___ 12. How many half-lifes must elapse for a radioisotope to contain one-sixteenth of its original mass?
(1) two
(2) three
(3) four
(4) five

___ 13. Which radioisotope is used in medicine?
(1) carbon-14
(2) cobalt-60
(3) boron-12
(4) actinium-225

___ 14. Exposure to radiation is measured in a unit called the
(1) positron.
(2) mass defect.
(3) half-life.
(4) rem.

ANSWERS TO SELF-CHECK QUESTIONS

- $_{-1}^{0}e$

- $_{90}^{234}Th \rightarrow {}_{91}^{234}Pa + {}_{-1}^{0}e$

- $_{84}^{212}Po \rightarrow {}_{82}^{208}Pb + {}_{2}^{4}He$

- $_{8}^{15}O \rightarrow {}_{7}^{15}N + {}_{+1}^{0}e$

- $_{92}^{235}U + {}_{1}^{0}n \rightarrow {}_{35}^{87}Br + {}_{57}^{146}La + 2{}_{0}^{1}n$

- The difference in mass is called the mass defect because the products have a total mass that is less than the total mass of the reactants.

- Its half life is 30.23 years. The following reaction shows how it decays: $_{55}^{137}Cs \rightarrow {}_{-1}^{0}e + {}_{56}^{137}Ba$

- 1 milligram

- Your answer should agree with the value listed in *Reference Table N*, which is 14.3 days.

- A person may experience nausea, loss of hair, ulcers, and internal bleeding.

- $_{95}^{241}Am \rightarrow {}_{93}^{237}Np \; {}_{+2}^{4}He$

Name ______________________ Class ______________ Date __________

UNIT XV

Holt Chemistry: The Physical Setting

Questions for Regents Practice

Focus On The Regents Exam

Nuclear Chemistry

PART A

_____ **1.** Which particle emitted in a nuclear decay has mass but no charge?
(a) alpha particle
(b) beta particle
(c) gamma ray
(d) neutron

_____ **2.** What is the term that is applied to the process in which an atom is bombarded with a high-energy particle to create a different element?
(a) natural transmutation
(b) artificial transmutation
(c) fission
(d) radioactive decay

_____ **3.** As a sample of radioactive cobalt-60 decays over time, its half-life
(a) remains constant.
(b) increases.
(c) decreases.
(d) increases at times, and decreases at other time.

_____ **4.** Which process represents natural nuclear decay?
(a) $^{235}_{92}U + ^{1}_{0}n \longrightarrow ^{87}_{35}Br + ^{146}_{57}La + 2^{1}_{0}n$
(b) $^{38}_{19}K \longrightarrow ^{38}_{20}Ca + ^{0}_{-1}e$
(c) $^{21}_{10}Ne + ^{4}_{2}He \longrightarrow ^{24}_{12}Mg + ^{1}_{0}n$
(d) $KClO_3 \longrightarrow K^+ + ClO_3^-$

_____ **5.** Which type of nuclear emission has the highest penetrating power?
(a) alpha
(b) beta
(c) gamma
(d) neutron

_____ **6.** Which radioisotope emits a beta particle when it decays?
(a) calcium-37
(b) francium-220
(c) plutonium-239
(d) strontium-90

_____ **7.** Transmutation occurs
(a) spontaneously, only.
(b) by artificial means, only.
(c) either spontaneously or artificially.
(d) only when a radioisotope emits either an alpha or beta particle.

_____ **8.** Alpha particle and beta particles differ in
(a) mass, only.
(b) charge, only.
(c) both mass and charge.
(d) neither mass nor charge.

_____ **9.** Examine the following reaction.

$$^{208}_{82}Pb + ^{58}_{26}Fe \longrightarrow ^{265}_{108}Hs + ^{1}_{0}n$$

The reaction above is an example of a(n)
(a) natural transmutation.
(b) artificial transmutation.
(c) fission reaction.
(d) fusion reaction.

Unit XV Nuclear Chemistry continued

_____ 10. Americium-241 can be safely used in smoke detectors because it
  (a) emits alpha particles when it decays.
  (b) does not emit any radioactive particles.
  (c) emits gamma radiation that ionizes the air molecules.
  (d) decays to form isotopes that are not radioactive.

PART B-1

_____ 11. Which nuclear equation is correctly balanced?
  (a) $^{14}_{7}N + ^{4}_{2}He \rightarrow ^{17}_{8}O + ^{2}_{1}H$
  (b) $^{254}_{99}Es + ^{4}_{2}He \rightarrow ^{258}_{101}Md + 2^{1}_{0}n$
  (c) $^{6}_{3}Li + 2^{1}_{0}n \rightarrow ^{4}_{2}He + ^{3}_{1}H$
  (d) $^{37}_{18}Ar + ^{0}_{-1}e \rightarrow ^{37}_{17}Cl$

_____ 12. How much time must elapse before 20 grams of nitrogen-16 decays and leaves 2.5 grams of the original radioisotope?
  (a) 2×7.2 seconds
  (b) 3×7.2 seconds
  (c) 4×7.2 seconds
  (d) 3×14.3 days

_____ 13. Examine the following reaction.

$$^{253}_{99}Es + ^{4}_{2}He \rightarrow ^{1}_{0}n + X$$

What does X represent?
  (a) $^{256}_{97}Bk$
  (b) $^{256}_{101}Md$
  (c) $^{257}_{101}Md$
  (d) $^{256}_{102}No$

_____ 14. Which statement explains why nuclear wastes may pose a problem?
  (a) They have short half-lifes and decay quickly.
  (b) They have short half-lifes and decay very slowly.
  (c) They have long half-lifes and decay quickly.
  (d) They have long half-lifes and decay very slowly.

_____ 15. According to *Reference Table N*, which radioisotope would be best to use for dating the age of Earth?
  (a) carbon-14
  (b) cobalt-60
  (c) thorium-232
  (d) radon-222

Unit XV Nuclear Chemistry continued

PART B-2

16. Write the balanced nuclear equation that shows the decay of radium-226.

17. A sample contains 3.5 grams of strontium-90. How many grams of this radioisotope were present in the sample 140.5 years ago?

18. The cobalt-60 used to treat certain types of cancer is made by bombarding cobalt-59 with a neutron.

(a) Write the balanced equation for this nuclear reaction.

(b) Is this reaction an example of a transmutation? Explain your answer.

PART C

Alpha particles were used to construct the model of the atom. In these experiments, alpha particles were aimed at a thin layer of gold atoms. Most of the alpha particles passed right through the gold foil. This observation led scientists to suggest that an atom consists mostly of empty space. Some of the alpha particles were deflected straight back, including ones that were deflected back toward their source. An illustration of this experiment is shown below.

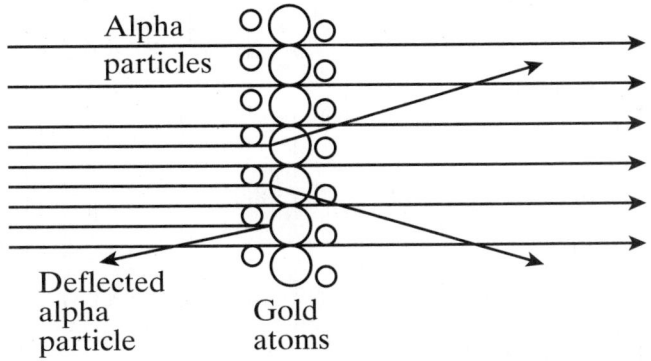

19. Explain why some of the alpha particles were deflected back toward their source.

Unit XV Nuclear Chemistry continued

20. Explain why this experiment could be performed without too much concern about the dangers of being exposed to radioisotopes.

Name _____ Class _____ Date _____

Holt Chemistry: The Physical Setting

Measurements and Calculations

Chemistry is both a qualitative and quantitative science. Matter can be described as rough or smooth, light or heavy, and solid or liquid. These properties of matter are described in qualitative terms. Whenever possible, however, scientists prefer to describe matter in quantitative terms, that is, with numbers.

SI Units

The numbers used in chemistry are expressed in SI units. There are seven SI base units.

Table 2 SI Base Units

Quantity	Symbol	Unit	Abbreviation
Length	l	meter	m
Mass	m	kilogram	kg
Time	t	second	s
Thermodynamic temperature	T	kelvin	K
Amount of substance	n	mole	mol
Electric current	I	ampere	A
Luminous intensity	I_v	candela	cd

When working with numbers, be sure to distinguish between quantity and unit. A **quantity** is something that has magnitude, size, or volume. A **unit** is a quantity adopted as a standard of measurement. For example, the mass of a solid can be reported as 2.4 grams. The *2.4*, or the measured mass, represents the quantity, while *grams* represent the unit in which the measured mass is reported.

You may have noticed in the table shown above that volume is not a quantity with a SI base unit. Volume is a quantity that needs a unit other than one of the seven basic SI units. Multiplying or dividing the base units derives units that are needed. For example, volume can be found by multiplying length × width × height. So, the unit of volume is the cubic meter, m^3. Because this unit is too large for practical use in a laboratory, scientists use the liter (L) as the unit for volume.

What You'll Need to Learn

This topic is part of the Regents Curriculum for the Physical Setting Exam.
Standard 1 (Mathematical Analysis), Performance Indicator 1: *Abstraction and symbolic representation are used to communicate mathematically.*

What Terms You'll Need to Know

accuracy
meniscus
precision
quantity
scientific notation
significant figure
unit

Where You Can Learn Even More:

Holt Chemistry: The Physical Setting
Chapter 1: The Science of Chemistry, Section 2
Chapter 2: Matter and Energy, Section 3

Measurements and Calculations continued

One liter is defined as one-thousandth of a cubic meter. In other words, 1 L = 1000 cm^3, which can also be expressed as follows.

$$1 \text{ L} = 1000 \text{ cm}^3 = 1000 \text{ mL}$$

In some cases, base units can be too small or too large for some measurements. For example, reporting the volume of a liquid to add to a test tube in liters would not be appropriate. When the base units are either too small or too large, a prefix can be added to make the unit either smaller or larger. *Reference Table C* has a list of selected prefixes. For example, the milliliter is the most appropriate unit for reporting the volume of a liquid that is to be poured into a test tube.

In chemistry, a measurement must often be converted from one unit to another. For example, you may want to convert 0.25 grams to milligrams. *Reference Table C* indicates that the prefix *milli-* represents a factor of 10^{-3}.

Therefore, 1 mg = 10^{-3} g or 0.001 g. This conversion can also be written as 1 g = 10^3 mg or 1000 mg. The conversion factor then is written as follows.

$$\frac{1 \text{ g}}{1000 \text{ mg}} \text{ or } \frac{1000 \text{ mg}}{1 \text{ g}}$$

The following shows how to convert 0.25 g to mg.

$$0.25 \text{ g} \times \frac{1000 \text{ mg}}{1 \text{ g}} = 250 \text{ mg}$$

Use the conversion factor so that the units cancel to leave the unit required in the answer. In the example above, the conversion factor needed must cancel grams and leave milligrams.

Self-Check Convert 804 milliliters (mL) to liters (L).

Making Measurements

Any measurement must be made with the right piece of equipment. Measurements of mass and length are less of a concern than measurements of volume. Mass is measured on a balance, while length is measured with a ruler. However, several pieces of glassware can be used to measure volume. These include beakers, burets, and graduated cylinders. In addition, these pieces of glassware come in different sizes. Which piece of glassware would you use to measure 1.4 mL?

Measuring a volume of 1.4 mL is best done with a buret, which is calibrated to the nearest 0.1 mL. However, if a procedure calls for a volume of 140 mL, then a graduated cylinder is

Measurements and Calculations *continued*

the best piece of glassware to use. A beaker should not be used for measuring a volume, even if the volume is 1.4 L. In this case, a 1 L-graduated cylinder is the best piece of glassware to use.

When using a graduated cylinder, be sure to read the volume correctly. A liquid forms a curved surface, known as a **meniscus**, inside a graduated cylinder as shown below.

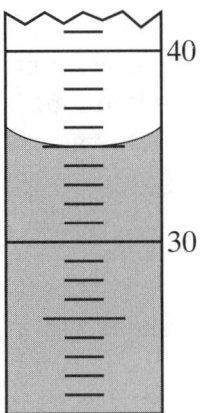

The measurement is always taken as the volume that corresponds to the bottom of the meniscus. Notice in the illustration above, that the liquid inside the graduated cylinder reaches a volume of 36 mL. However, the measurement is the value recorded at the bottom of the meniscus. Therefore, the volume in the illustration above should be recorded as 35 mL.

Self-Check What is the volume in the graduated cylinder shown below?

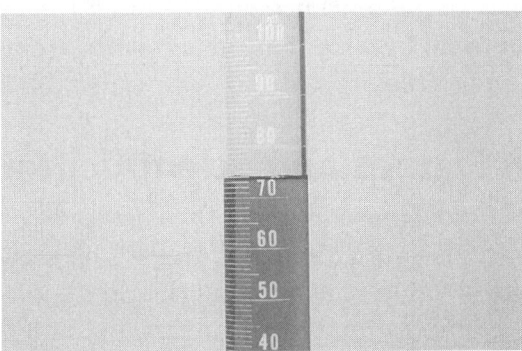

Using the proper equipment and glassware reduces the chances of introducing an error when making a measurement and therefore increases the **accuracy** of the result. Accuracy is a description of how close a measurement is to the true value of the quantity measured.

Notes/Study Ideas/Answers

For example, suppose that the true value of a volume of a liquid is 9.4 mL. Measuring this volume in a 100-mL graduated cylinder will not give you the same accuracy that measuring the volume in a 10-mL graduated cylinder will give you.

If you pour the liquid into the 10-mL graduated cylinder, you should be able to read the volume as 9.4 mL. However, if you pour the liquid into a 100-mL graduated cylinder, you will be able to record a volume of 9 mL. You can only approximate the decimal value by estimating how much the bottom of the meniscus extends above the 9 mL mark. You may record the volume as 9.5 mL, a value that is not as accurate as the one obtained by using a 10-mL graduated cylinder.

In addition to accuracy, scientists also consider **precision** when making measurements. Precision refers to the exactness of a measurement, or how closely several measurements of the same quantity made in the same way agree with one another.

Self-Check Why do the darts shown below reflect low accuracy but high precision?

Accuracy and precision are two reasons why scientists do not depend on a single measurement. Rather, they make several measurements to make sure that their results are as accurate and precise as possible.

Significant Figures and Scientific Notation

When reporting measurements and the results of calculations, scientists always use **significant figures**. The significant figures in a reported value consist of all the digits known with certainty as well as one estimated, or uncertain, digit. Notice that that the term "significant" does not mean "certain." The last digit is uncertain or estimated.

For example, suppose you use a thermometer calibrated in one-degree increments to record a temperature of a solution in a test tube. You record the temperature as 22°C. The two digits in your recorded value are both significant figures. The first one is known with certainty, while the second digit is an estimated

Measurements and Calculations *continued*

value. The actual temperature of the solution is between 21°C and 23°C. You estimate the temperature to be 22°C. Notice that the actual value has a range of two degrees.

Now suppose you use a thermometer that is calibrated in one-tenth degree increments. This time when you measure the temperature of the solution, you record a reading of 21.6°C. All three digits in your reported value are significant. The first two digits are known with certainty, while the last digit is estimated. You know the temperature is between 21.0°C and 22.0°C, and you estimate it to be 21.6°C. Notice that the actual value has a range of one degree.

There are several rules for determining whether a digit is significant.

1. *Nonzero digits are always significant.* For example, 14.54 g has four significant figures, while 528 km has three significant figures.

2. *Zeros between nonzero digits are significant.* For example, 40.7 mL has three significant figures, while 20,455 m has five significant figures.

3. *Zeros both at the end of a number and to the right of a decimal point are significant.* For example, 64.00 g has four significant figures, while 17.050 g has five significant figures.

4. *A decimal point placed after zeros indicates that the zeros are significant.* For example, 200. mL has three significant figures, while 4050. m has four significant figures.

5. *Zeros in front of nonzero digits are not significant, either on the left or right of a decimal point.* For example, 0.005 cm has one significant figure, while 0.0155 mg has three significant figures.

> **Self-Check** How many significant figures are there in each of the following values?
> (a) 0.0025 g
> (b) 54,209 km
> (c) 1.0001 cm
> (d) 400.0 mL

When you use a calculator to find a result, you must pay special attention to significant figures because calculators do not identify significant figures. For example, if you use your calculator to find the density of a sample by dividing its mass, 2.54 g, by its volume, 6.2 cm^3, your calculator may give you a value of 0.4096774 g/cm^3.

Notes/Study Ideas/Answers

Measurements and Calculations *continued*

Notes/Study Ideas/Answers

The mass is reported to three significant figures, while the volume is reported to two significant figures. However, the calculated density is reported to seven significant figures. The density cannot be reported to more than two significant figures, which is the same number in the value with the smallest number of significant figures. Therefore, the density is reported to the correct number of significant figures as 0.41 g/cm^3.

There are two rules for using significant figures in calculations.

1. *When multiplying or dividing numbers, the answer cannot have more significant figures than there are in the measurement with the smallest number of significant figures.* For example, when you multiply 12.506 m by 4.35 m, the answer can have only three significant figures.

2. *When adding or subtracting numbers, the answer cannot have more significant figures to the right of the decimal point than there are in the measurement with the smallest number of significant figures to the right of the decimal.* For example, if you add 3.97 g + 22.705 g + 184.5 g, then the answer can have only one digit to the right of the decimal point. In this case, the answer will have four significant figures, whereas one of the values, 3.97 g, has only three significant figures.

> **Self-Check** Perform the following calculations and express your answer to the correct number of significant figures.
> (a) 3.05 g/8.472 mL
> (b) (26.02 m)(1.04 m)
> (c) 1.0001 cm − 0.012 cm
> (d) 32.89 mm + 14.2 mm

When making measurements and performing calculations, chemists often work with very small and very large numbers. For example, the number of atoms of iron in 1 mol Fe is 602,213,670,000,000,000,000,000 atoms. Writing this number can be tedious. Fortunately, there is another way to write very small and very large numbers that takes far less time. This is **scientific notation**.

For example, the number above is written in scientific notation as 6.0221367 x 10^{23}. Scientific notation is written so that only one digit is placed to the left of the decimal point. As you can see, scientific notation also uses powers of ten. The 10^{23} means that the decimal point must be moved twenty-three places to the right to write out all the digits in this value.

The exponents in scientific notation can also have negative value. For example, a scientist may report a mass as having a

Measurements and Calculations *continued*

value of 2.4×10^{-3} kg. In this case, the 10^{-3} means that the decimal point must be moved three places to the left to write out all the digits in this value—0.0024 kg. Notice that if you wanted to express 0.0024 kg in scientific notation, you must move the decimal point three places to the right so that only one digit appears to the left of the decimal point. If the decimal is moved to the right, the exponent is negative. Therefore, 0.0024 kg is written as 2.4×10^{-3} kg.

Self-Check Rewrite the following values in scientific notation.
(a) 0.000673 g
(b) 504,986,000 km
(c) 100.01 cm

ANSWERS TO SELF-CHECK QUESTIONS

- 0.804 mL

- 73 mL

- They are clustered in one area so they show high precision, but they are far from the bull's-eye so they show poor accuracy.

- (a) two significant figures
  (b) five significant figures
  (c) five significant figures
  (d) four significant figures

- (a) 0.360 g/mL
  (b) 27.1 m^2
  (c) 0.988 cm
  (d) 47.1 mm

- (a) 6.73×10^{-4} g
  (b) 5.04986×10^8 km
  (c) 1.0001×10^2 cm

Holt Chemistry: The Physical Setting

Reference Tables

Table A
Standard Temperature and Pressure

Name	Value	Unit
Standard Pressure	101.3 kPa 1 atm	kilopascal atmosphere
Standard Temperature	273 K 0°C	kelvin degree Celsius

Notice that standard pressure is reported in two units (kilopascal and atmosphere), and standard temperature is also reported in two units (Kelvin and degree Celsius). Table A is helpful when working with the gas laws. This table also has the information needed to change degrees Celsius into kelvins, the unit that you must use when solving problems that deal with Charles's law.

You will be provided with Reference Tables A-T and a periodic table with the Regents Chemistry examination. Many of the questions on the examination will require the use of these tables. In fact, some questions can be answered by simply checking the appropriate reference table. Therefore, it is important that you are familiar with the information that each table contains and where you are likely to need that information.

Self-Check A gas occupies 4.5 L at standard temperature and pressure. What will the volume be if the temperature is increased to 42°C at constant pressure?

Table B
Physical Constants for Water

Heat of Fusion	333.6 J/g
Heat of Vaporization	2259 J/g
Specific Heat Capacity of $H_2O\ (\ell)$	4.2 J/g•K

Use the heat of fusion to calculate how much energy is required to change a given mass of ice to water. Use the heat of vaporization to calculate how much energy is needed to change a given mass of water to vapor.

Self-Check How much energy is needed to melt 4.5 g of ice? How much energy is needed to vaporize 25.5 g of water?

Reference Tables continued

Table C
Selected Prefixes

Factor	Prefix	Symbol
10^3	kilo-	k
10^{-1}	deci-	d
10^{-2}	centi-	c
10^{-3}	milli-	m
10^{-6}	micro-	μ
10^{-9}	nano-	n
10^{-12}	pico-	p

These prefixes indicate how to convert in the metric system. For example, the prefix *micro* indicates a factor of 10^{-6}. Therefore, you move the decimal point six places to the left to convert micrograms into grams. Conversely, you move the decimal point six places to the right to convert grams into micrograms.

Self-Check Convert 4.5 cm to mm and convert 582 mL to L.

Table D
Selected Units

Symbol	Name	Quantity
m	meter	length
kg	kilogram	mass
Pa	pascal	pressure
K	kelvin	temperature
mol	mole	amount of substance
J	joule	energy, work, quantity of heat
s	second	time
L	liter	volume
ppm	part per million	concentration
M	molarity	solution concentration

Reference Tables continued

Several SI base units are listed for a variety of measured quantities common in chemistry and related sciences.

Table E
Selected Polyatomic Ions

Formula	Name	Formula	Name
H_3O^+	hydronium	CrO_4^{2-}	chromate
Hg_2^{2+}	dimercury (I)	$Cr_2O_7^{2-}$	dichromate
NH_4^+	ammonium	MnO_4^-	permanganate
$C_2H_3O_2^-$ / CH_3COO^-	acetate	NO_2^-	nitrite
		NO_3^-	nitrate
CN^-	cyanide	O_2^{2-}	peroxide
CO_3^{2-}	carbonate	OH^-	hydroxide
HCO_3^-	hydrogen carbonate	PO_4^{3-}	phosphate
$C_2O_4^{2-}$	oxalate	SCN^-	thiocyanate
ClO^-	hypochlorite	SO_3^{2-}	sulfite
ClO_2^-	chlorite	SO_4^{2-}	sulfate
ClO_3^-	chlorate	HSO_4^-	hydrogen sulfate
ClO_4^-	perchlorate	$S_2O_3^{2-}$	thiosulfate

This table is useful for writing the formulas for ionic compounds that contain a polyatomic ion. For example, the formula for copper(II) phosphate is $Cu_3(PO_4)_2$. Remember to enclose a polyatomic ion in parentheses if more than one unit is present in the compound. Also notice that this table contains the formula for the hydronium ion, H_3O^+, that is characteristic of acids.

Table F
Solubility Guidelines

Ions That Form Soluble Compounds	Exceptions
Group 1 ions (Li^+, Na^+, etc.)	
ammonium (NH_4^+)	
nitrate (NO_3^-)	
acetate ($C_2H_3O_2^-$ or CH_3COO^-)	
hydrogen carbonate (HCO_3^-)	
chlorate (ClO_3^-)	
perchlorate (ClO_4^-)	
halides (Cl^-, Br^-, I^-)	when combined with Ag^+, Pb^{2+}, and Hg_2^{2+}
sulfates (SO_4^{2-})	when combined with Ag^+, Ca^{2+}, Sr^{2+}, Ba^{2+}, and Pb^{2+}

Ions That Form Insoluble Compounds	Exceptions
carbonate (CO_3^{2-})	when combined with Group 1 ions or ammonium (NH_4^+)
chromate (CrO_4^{2-})	when combined with Group 1 ions or ammonium (NH_4^+)
phosphate (PO_4^{3-})	when combined with Group 1 ions or ammonium (NH_4^+)
sulfide (S^{2-})	when combined with Group 1 ions or ammonium (NH_4^+)
hydroxide (OH^-)	when combined with Group 1 ions, Ca^{2+}, Ba^{2+}, or Sr^{2+}

Table F at the top lists the ions that form *soluble* compounds, while Table F beneath lists the ions that form *insoluble* compounds. Note the exceptions in both tables. The information in these tables is used to determine if an ionic compound is soluble. For example, calcium chlorate, $Ca(ClO_3)_2$, is soluble, whereas calcium carbonate, $CaCO_3$, is insoluble.

Self-Check Determine if the following compounds are soluble or insoluble: ammonium chloride, lead sulfate, magnesium sulfide, and calcium chromate.

Reference Tables *continued*

Table G Solubility Curves

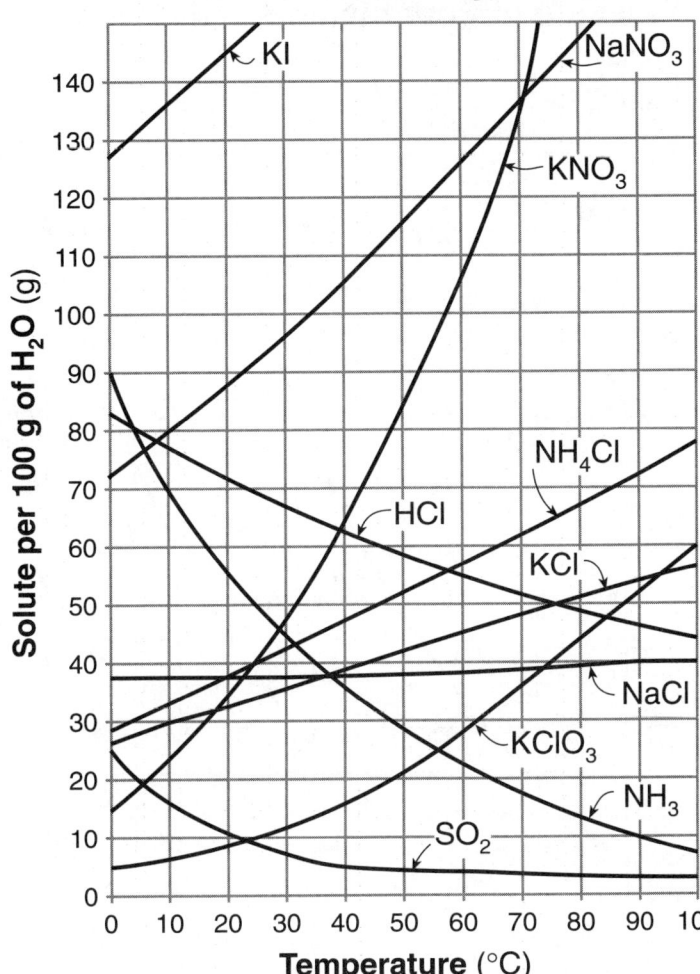

This table provides information about solutes and solutions. The line for each compound shows how many grams of a solute can dissolve in 100 grams of water at a given temperature. For example, 20 g of $KClO_3$ can dissolve in 100 g of water at 48°C. This represents a saturated solution at this temperature. An unsaturated solution would contain less than 20 g of $KClO_3$, while a supersaturated solution would contain more than 20 g of $KClO_3$ at 48°C. Notice that solubility generally increases as the temperature increases. However, some compounds, such as SO_2, show a decreased solubility as the temperature increases.

Self-Check At 60°C, a solution contains 35 g of KCl. Is this solution unsaturated, saturated, or supersaturated? Explain your answer.

Reference Tables continued

Table H
Vapor Pressure of Four Liquids

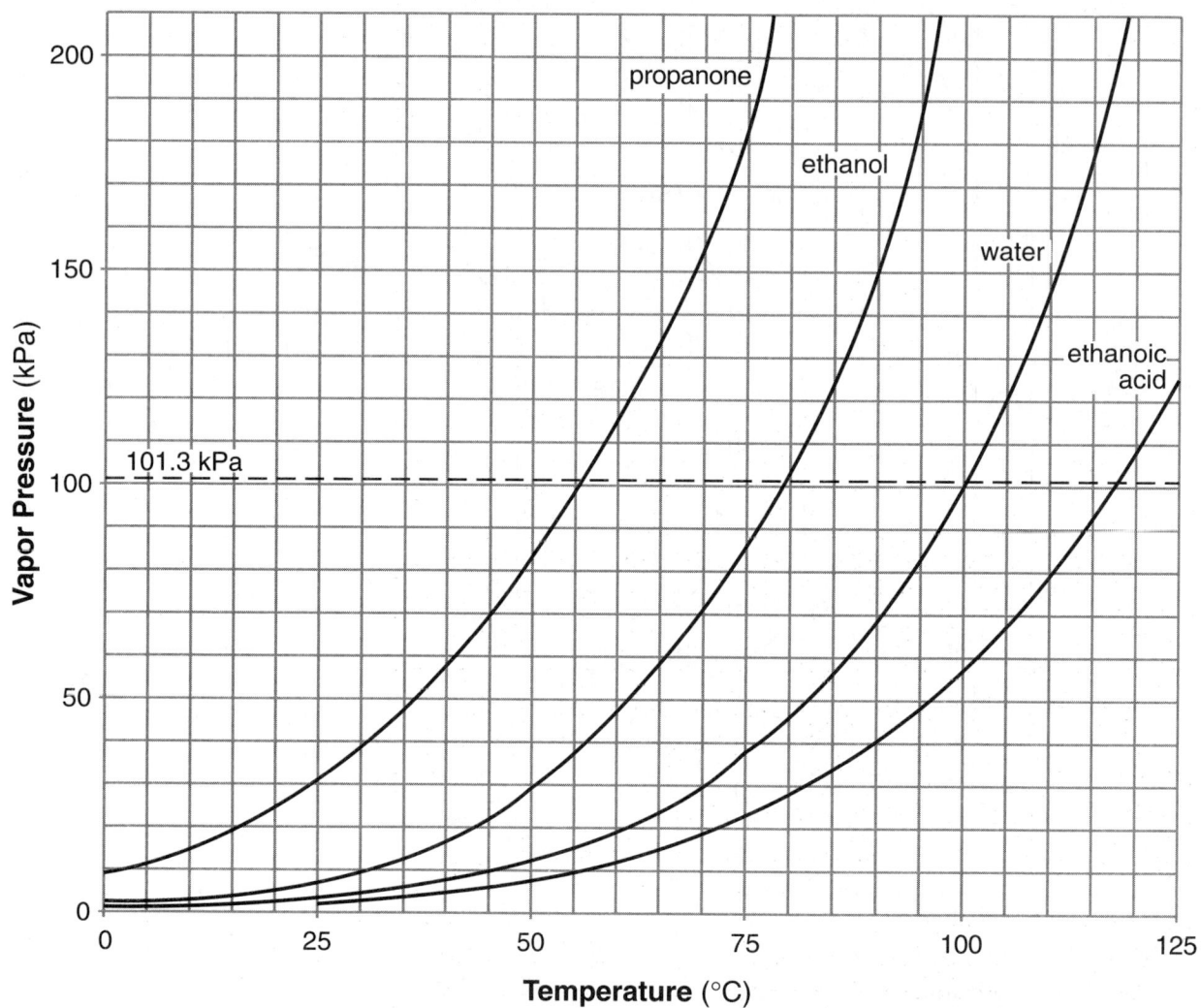

Notes/Study Ideas/Answers

Notice that the vapor pressure for four liquids can be determined from the graph in this table. For example, at 50°C, ethanol exerts a vapor pressure of 30 kPa. The dotted line represents standard pressure, 101.3 kPa. This graph can also be used to determine the boiling point of each liquid at standard pressure. The boiling point is reached when the vapor pressure of a substance equals atmospheric pressure. Therefore, the boiling point of ethanol is about 80°C.

Self-Check What is the vapor pressure of propanone at 40°C? What is the boiling point of propanone?

Reference Tables continued

Table I
Heats of Reaction at 101.3 kPa and 298 K

Reaction	ΔH (kJ)*
$CH_4(g) + 2O_2(g) \longrightarrow CO_2(g) + 2H_2O(\ell)$	−890.4
$C_3H_8(g) + 5O_2(g) \longrightarrow 3CO_2(g) + 4H_2O(\ell)$	−2219.2
$2C_8H_{18}(\ell) + 25O_2(g) \longrightarrow 16CO_2(g) + 18H_2O(\ell)$	−10943
$2CH_3OH(\ell) + 3O_2(g) \longrightarrow 2CO_2(g) + 4H_2O(\ell)$	−1452
$C_2H_5OH(\ell) + 3O_2(g) \longrightarrow 2CO_2(g) + 3H_2O(\ell)$	−1367
$C_6H_{12}O_6(s) + 6O_2(g) \longrightarrow 6CO_2(g) + 6H_2O(\ell)$	−2804
$2CO(g) + O_2(g) \longrightarrow 2CO_2(g)$	−566.0
$C(s) + O_2(g) \longrightarrow CO_2(g)$	−393.5
$4Al(s) + 3O_2(g) \longrightarrow 2Al_2O_3(s)$	−3351
$N_2(g) + O_2(g) \longrightarrow 2NO(g)$	+182.6
$N_2(g) + 2O_2(g) \longrightarrow 2NO_2(g)$	+66.4
$2H_2(g) + O_2(g) \longrightarrow 2H_2O(g)$	−483.6
$2H_2(g) + O_2(g) \longrightarrow 2H_2O(\ell)$	−571.6
$N_2(g) + 3H_2(g) \longrightarrow 2NH_3(g)$	−91.8
$2C(s) + 3H_2(g) \longrightarrow C_2H_6(g)$	−84.0
$2C(s) + 2H_2(g) \longrightarrow C_2H_4(g)$	+52.4
$2C(s) + H_2(g) \longrightarrow C_2H_2(g)$	+227.4
$H_2(g) + I_2(g) \longrightarrow 2HI(g)$	+53.0
$KNO_3(s) \xrightarrow{H_2O} K^+(aq) + NO_3^-(aq)$	+34.89
$NaOH(s) \xrightarrow{H_2O} Na^+(aq) + OH^-(aq)$	−44.51
$NH_4Cl(s) \xrightarrow{H_2O} NH_4^+(aq) + Cl^-(aq)$	+14.78
$NH_4NO_3(s) \xrightarrow{H_2O} NH_4^+(aq) + NO_3^-(aq)$	+25.69
$NaCl(s) \xrightarrow{H_2O} Na^+(aq) + Cl^-(aq)$	+3.88
$LiBr(s) \xrightarrow{H_2O} Li^+(aq) + Br^-(aq)$	−48.83
$H^+(aq) + OH^-(aq) \longrightarrow H_2O(\ell)$	−55.8

*Minus sign indicates an exothermic reaction.

This table provides data on the quantity of energy that is either absorbed or released as heat for each reaction that is listed. A positive ΔH value indicates that the reaction is endothermic, while a negative ΔH value indicates an exothermic reaction. For example, when the reaction $2H_2(g) + O_2(g) \rightarrow 2H_2O(g)$ occurs, 483.6 kJ of energy as heat is released.

Notes/Study Ideas/Answers

Name _____ Class _____ Date _____

Reference Tables continued

Notes/Study Ideas/Answers

Self-Check What is the energy change that occurs when the following reaction takes place?
$2C(s) + H_2(g) \longrightarrow C_2H_2(g)$
Is the reaction above endothermic or exothermic? Explain your answer.

Table J
Activity Series**

Most → Least	Metals	Nonmetals	Most → Least
	Li	F_2	
	Rb	Cl_2	
	K	Br_2	
	Cs	I_2	
	Ba		
	Sr		
	Ca		
	Na		
	Mg		
	Al		
	Ti		
	Mn		
	Zn		
	Cr		
	Fe		
	Co		
	Ni		
	Sn		
	Pb		
	**H_2		
	Cu		
	Ag		
	Au		

**Activity Series based on hydrogen standard

Reference Tables continued

Notice that metals are listed on the left side, and nonmetals on the right side. As you proceed down the list, the activity of the elements decreases. Li is the most active metal, while Au is the least active metal. F_2 is the most active nonmetal. This table is used to predict whether a single-replacement reaction occurs spontaneously and how electrons flow in an electrochemical cell. The higher on the table an element is, the more likely it is to undergo oxidation (lose electrons).

For example, adding Na(s) to $Zn(Cl)_2(aq)$ will bring about a spontaneous reaction because Na is more active than Zn and will replace it to form NaCl(aq). Assume that one electrode in an electrochemical cell is Mg, and the other electrode is Fe. Electrons will flow from Mg to Fe because Mg loses electrons more readily, as shown by Mg being above Fe on the table.

Self-Check Will adding Fe(s) to $CaSO_4(aq)$ result in a spontaneous reaction? Explain your answer. In an electrochemical cell, which electrode will act as the anode—Ca or Ni? Explain your answer.

Table K
Common Acids

Formula	Name
HCl(aq)	hydrochloric acid
HNO_3(aq)	nitric acid
H_2SO_4(aq)	sulfuric acid
H_3PO_4(aq)	phosphoric acid
H_2CO_3(aq) or CO_2(aq)	carbonic acid
CH_3COOH(aq) or $HC_2H_3O_2$(aq)	ethanoic acid (acetic acid)

Reference Tables *continued*

Table L
Common Bases

Formula	Name
NaOH(aq)	sodium hydroxide
KOH(aq)	potassium hydroxide
Ca(OH)$_2$(aq)	calcium hydroxide
NH$_3$(aq)	aqueous ammonia

Table M
Common Acid–Base Indicators

Indicator	Approximate pH Range for Color Change	Color Change
methyl orange	3.2–4.4	red to yellow
bromthymol blue	6.0–7.6	yellow to blue
phenolphthalein	8.2–10	colorless to pink
litmus	5.5–8.2	red to blue
bromcresol green	3.8–5.4	yellow to blue
thymol blue	8.0–9.6	yellow to blue

An Arrhenius acid produces H$^+$ or H$_3$O$^+$ ions in solution. An Arrhenius base produces OH$^-$ ions in solution. A strong acid reacts with a strong base to produce a salt and water with a pH value of 7. Notice that Table M lists common indicators and their pH range for color change. For example, bromthymol blue changes from yellow to blue as the pH changes from 6.0 to 7.6

Self-Check Write a balanced equation to show the neutralization reaction that occurs between HNO$_3$ and Ca(OH)$_2$. Would methyl orange be an appropriate indicator to use in this titration? Explain your answer.

Table N
Selected Radioisotopes

Nuclide	Half-Life	Decay Mode	Nuclide Name
^{198}Au	2.69 d	β^-	gold-198
^{14}C	5730 y	β^-	carbon-14
^{37}Ca	175 ms	β^+	calcium-37
^{60}Co	5.26 y	β^-	cobalt-60
^{137}Cs	30.23 y	β^-	cesium-137
^{53}Fe	8.51 min	β^+	iron-53
^{220}Fr	27.5 s	α	francium-220
^{3}H	12.26 y	β^-	hydrogen-3
^{131}I	8.07 d	β^-	iodine-131
^{37}K	1.23 s	β^+	potassium-37
^{42}K	12.4 h	β^-	potassium-42
^{85}Kr	10.76 y	β^-	krypton-85
^{16}N	7.2 s	β^-	nitrogen-16
^{19}Ne	17.2 s	β^+	neon-19
^{32}P	14.3 d	β^-	phosphorus-32
^{239}Pu	2.44×10^4 y	α	plutonium-239
^{226}Ra	1600 y	α	radium-226
^{222}Rn	3.82 d	α	radon-222
^{90}Sr	28.1 y	β^-	strontium-90
^{99}Tc	2.13×10^5 y	β^-	technetium-99
^{232}Th	1.4×10^{10} y	α	thorium-232
^{233}U	1.62×10^5 y	α	uranium-233
^{235}U	7.1×10^8 y	α	uranium-235
^{238}U	4.51×10^9 y	α	uranium-238

ms = milliseconds; s = seconds; min = minutes;
h = hours; d = days; y = years

Reference Tables *continued*

Table O
Symbols Used in Nuclear Chemistry

Name	Notation	Symbol
alpha particle	$^{4}_{2}He$ or $^{4}_{2}\alpha$	α
beta particle (electron)	$^{0}_{-1}e$ or $^{0}_{-1}\beta$	β^{-}
gamma radiation	$^{0}_{0}\gamma$	γ
neutron	$^{1}_{0}n$	n
proton	$^{1}_{1}H$ or $^{1}_{1}p$	p
positron	$^{0}_{+1}e$ or $^{0}_{+1}\beta$	β^{+}

Table N on the previous page lists selected radioisotopes, atomic symbols, half-lives, and decay modes. For example, krypton-85, ^{85}Kr, has a half-life of 10.76 years, and decays by emitting a beta particle. The symbols for radioactive particles are listed in Table O. These two tables enable you to write balanced nuclear equations and to solve problems involving half-lives.

Self-Check Write the nuclear equation that shows the decay of francium-220. If a sample contains 20 g of francium-220, how much of this radioisotope will remain after 1 minute and 50 seconds?

Table P
Organic Prefixes

Prefix	Number of Carbon Atoms
meth-	1
eth-	2
prop-	3
but-	4
pent-	5
hex-	6
hept-	7
oct-	8
non-	9
dec-	10

Reference Tables continued

Table Q
Homologous Series of Hydrocarbons

Name	General Formula	Examples	
		Name	Structural Formula
alkanes	C_nH_{2n+2}	ethane	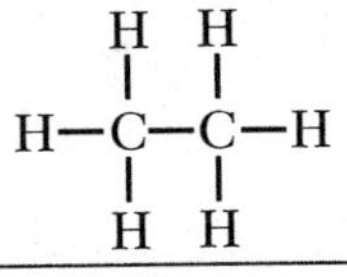 H—C(H)(H)—C(H)(H)—H
alkenes	C_nH_{2n}	ethene	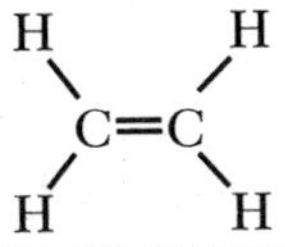 H₂C=CH₂
alkynes	C_nH_{2n-2}	ethyne	H—C≡C—H

n = number of carbon atoms

Check Table P if you need to know the number of carbon atoms in an organic compound. For example, the prefix *oct-* indicates that both octane and octanol contain eight carbon atoms. More information about hydrocarbons is provided in Table Q. Using both tables, you should be able to write the formula for organic compounds such as pentene, C_5H_{10}, and heptyne, C_7H_{12}.

Self-Check Write the formulas for nonane and butyne. Which one is an unsaturated compound? Explain your answer.

Table R
Organic Functional Groups

Class of Compound	Functional Group	General Formula	Example
halide (halocarbon)	—F (fluoro-) —Cl (chloro-) —Br (bromo-) —I (iodo-)	R—X (X represents any halogen)	$CH_3CHClCH_3$ 2-chloropropane
alcohol	—OH	R—OH	$CH_3CH_2CH_2OH$ 1-propanol
ether	—O—	R—O—R'	$CH_3OCH_2CH_3$ methyl ethyl ether
aldehyde	$\overset{\overset{\displaystyle O}{\|\|}}{-C-H}$	$\overset{\overset{\displaystyle O}{\|\|}}{R-C-H}$	$CH_3CH_2\overset{\overset{\displaystyle O}{\|\|}}{C}-H$ propanal
ketone	$\overset{\overset{\displaystyle O}{\|\|}}{-C-}$	$\overset{\overset{\displaystyle O}{\|\|}}{R-C-R'}$	$CH_3\overset{\overset{\displaystyle O}{\|\|}}{C}CH_2CH_2CH_3$ 2-pentanone
organic acid	$\overset{\overset{\displaystyle O}{\|\|}}{-C-OH}$	$\overset{\overset{\displaystyle O}{\|\|}}{R-C-OH}$	$CH_3CH_2\overset{\overset{\displaystyle O}{\|\|}}{C}-OH$ propanoic acid
ester	$\overset{\overset{\displaystyle O}{\|\|}}{-C-O-}$	$\overset{\overset{\displaystyle O}{\|\|}}{R-C-O-R'}$	$CH_3CH_2\overset{\overset{\displaystyle O}{\|\|}}{C}OCH_3$ methyl propanoate
amine	$\overset{\displaystyle }{-N-}$	$\overset{\displaystyle R'}{R-N-R''}$	$CH_3CH_2CH_2NH_2$ 1-propanamine
amide	$\overset{\overset{\displaystyle O}{\|\|}}{-C-NH}$	$\overset{\overset{\displaystyle O}{\|\|}}{R-C-NH}\overset{\displaystyle R'}{}$	$CH_3CH_2\overset{\overset{\displaystyle O}{\|\|}}{C}-NH_2$ propanamide

R represents a bonded atom or group of atoms.

Reference Tables continued

The functional group determines the class to which an organic compound belongs. For example, the presence of the —OH group classifies the compound an alcohol. Use this table to determine the class of a compound, given its structural formula.

Self-Check To what class of organic compounds does the following molecule belong?

```
      H
      |
  H — C — Cl
      |
      H
```
chloromethane

Table S
Properties of Selected Elements

Atomic Number	Symbol	Name	Ionization Energy (kJ/mol)	Electro-negativity	Melting Point (K)	Boiling Point (K)	Density** (g/cm³)	Atomic Radius (pm)
1	H	hydrogen	1312	2.1	14	20	0.00009	208
2	He	helium	2372	—	1	4	0.000179	50
3	Li	lithium	520	1.0	454	1620	0.534	155
4	Be	beryllium	900	1.6	1551	3243	1.8477	112
5	B	boron	801	2.0	2573	3931	2.340	98
6	C	carbon	1086	2.6	3820	5100	3.513	91
7	N	nitrogen	1402	3.0	63	77	0.00125	92
8	O	oxygen	1314	3.4	55	90	0.001429	65
9	F	fluorine	1681	4.0	54	85	0.001696	57
10	Ne	neon	2081	—	24	27	0.0009	51
11	Na	sodium	496	0.9	371	1156	0.971	190
12	Mg	magnesium	736	1.3	922	1363	1.738	160
13	Al	aluminum	578	1.6	934	2740	2.698	143
14	Si	silicon	787	1.9	1683	2628	2.329	132
15	P	phosphorus	1012	2.2	44	553	1.820	128
16	S	sulfur	1000	2.6	386	718	2.070	127
17	Cl	chlorine	1251	3.2	172	239	0.003214	97
18	Ar	argon	1521	—	84	87	0.001783	88
19	K	potassium	419	0.8	337	1047	0.862	235
20	Ca	calcium	590	1.0	1112	1757	1.550	197
21	Sc	scandium	633	1.4	1814	3104	2.989	162
22	Ti	titanium	659	1.5	1933	3580	4.540	145
23	V	vanadium	651	1.6	2160	3650	6.100	134
24	Cr	chromium	653	1.7	2130	2945	7.190	130
25	Mn	manganese	717	1.6	1517	2235	7.440	135
26	Fe	iron	762	1.8	1808	3023	7.874	126
27	Co	cobalt	760	1.9	1768	3143	8.900	125
28	Ni	nickel	737	1.9	1726	3005	8.902	124
29	Cu	copper	745	1.9	1357	2840	8.960	128
30	Zn	zinc	906	1.7	693	1180	7.133	138
31	Ga	gallium	579	1.8	303	2676	5.907	141
32	Ge	germanium	762	2.0	1211	3103	5.323	137
33	As	arsenic	944	2.2	1090	889	5.780	139
34	Se	selenium	941	2.6	490	958	4.790	140
35	Br	bromine	1140	3.0	266	332	3.122	112
36	Kr	krypton	1351	—	117	121	0.00375	103
37	Rb	rubidium	403	0.8	312	961	1.532	248
38	Sr	strontium	549	1.0	1042	1657	2.540	215
39	Y	yttrium	600	1.2	1795	3611	4.469	178
40	Zr	zirconium	640	1.3	2125	4650	6.506	160

Reference Tables continued

Table S
Properties of Selected Elements continued

Atomic Number	Symbol	Name	Ionization Energy (kJ/mol)	Electro-negativity	Melting Point (K)	Boiling Point (K)	Density** (g/cm^3)	Atomic Radius (pm)
41	Nb	niobium	652	1.6	2741	5015	8.570	146
42	Mo	molybdenum	684	2.2	2890	4885	10.220	139
43	Tc	technetium	702	1.9	2445	5150	11.500	136
44	Ru	ruthenium	710	2.2	2583	4173	12.370	134
45	Rh	rhodium	720	2.3	2239	4000	12.410	134
46	Pd	palladium	804	2.2	1825	3413	12.020	137
47	Ag	silver	731	1.9	1235	2485	10.500	144
48	Cd	cadmium	868	1.7	594	1038	8.650	171
49	In	indium	558	1.8	429	2353	7.310	166
50	Sn	tin	709	2.0	505	2543	7.310	162
51	Sb	antimony	831	2.1	904	1908	6.691	159
52	Te	tellurium	869	2.1	723	1263	6.240	142
53	I	iodine	1008	2.7	387	458	4.930	132
54	Xe	xenon	1170	2.6	161	166	0.0059	124
55	Cs	cesium	376	0.8	302	952	1.873	267
56	Ba	barium	503	0.9	1002	1910	3.594	222
57	La	lanthanum	538	1.1	1194	3730	6.145	138
\multicolumn{9}{	c	}{Elements 58–71 have been omitted.}
72	Hf	hafnium	659	1.3	2503	5470	13.310	167
73	Ta	tantalum	728	1.5	3269	5698	16.654	149
74	W	tungsten	759	2.4	3680	5930	19.300	141
75	Re	rhenium	756	1.9	3453	5900	21.020	137
76	Os	osmium	814	2.2	3327	5300	22.590	135
77	Ir	iridium	865	2.2	2683	4403	22.560	136
78	Pt	platinum	864	2.3	2045	4100	21.450	139
79	Au	gold	890	2.5	1338	3080	19.320	146
80	Hg	mercury	1007	2.0	234	630	13.546	160
81	Tl	thallium	589	2.0	577	1730	11.850	171
82	Pb	lead	716	2.3	601	2013	11.350	175
83	Bi	bismuth	703	2.0	545	1833	9.747	170
84	Po	polonium	812	2.0	527	1235	9.320	167
85	At	astatine	—	2.2	575	610	—	145
86	Rn	radon	1037	—	202	211	0.00973	134
87	Fr	francium	393	0.7	300	950	—	270
88	Ra	radium	—	0.9	973	1413	5.000	233
89	Ac	actinium	499	1.1	1320	3470	10.060	—
\multicolumn{9}{	c	}{Elements 90 and above have been omitted.}

*Boiling point at standard pressure
**Density at STP

Use Table S if you need to check an element's symbol or atomic number. The ionization energy represents the energy required to remove a valence electron. The electronegativity value represents the attraction an atom has for the shared electrons in a bond. The difference in electronegativity values indicates the degree of polarity of a bond. This table also indicates how ionization energy, electronegativity, and atomic size change as you go across a period of down a group. You simply have to check the values for each element in a particular period or group.

Self-Check Which bond has a higher degree of polarity—H—O or H—S? Explain your answer.

Table T
Important Formulas and Equations

Density	$d = \dfrac{m}{V}$	d = density m = mass V = volume
Mole Calculations	number of moles = $\dfrac{\text{given mass (g)}}{\text{gram-formula mass}}$	
Percent Error	% error = $\dfrac{\text{measured value} - \text{accepted value}}{\text{accepted value}} \times 100$	
Percent Composition	% composition by mass = $\dfrac{\text{mass of part}}{\text{mass of whole}} \times 100$	
Concentration	parts per million = $\dfrac{\text{grams of solute}}{\text{grams of solution}} \times 1\,000\,000$	
	molarity = $\dfrac{\text{moles of solute}}{\text{liters of solution}}$	
Combined Gas Law	$\dfrac{P_1 V_1}{T_1} = \dfrac{P_2 V_2}{T_2}$	P = pressure V = volume T = temperature (K)
Titration	$M_A V_A = M_B V_B$	M_A = molarity of H$^+$ M_B = molarity of OH$^-$ V_A = volume of acid V_B = volume of base
Heat	$q = mC\Delta T$ $q = mH_f$ $q = mH_v$	q = heat H_f = heat of fusion m = mass H_v = heat of vaporization C = specific heat capacity ΔT = change in temperature
Temperature	K = °C + 273	K = kelvin °C = degrees Celsius
Radioactive Decay	fraction remaining = $\left(\dfrac{1}{2}\right)^{\frac{t}{T}}$ number of half-life periods = $\dfrac{t}{T}$	t = total time elapsed T = half-life

Check this table for a formula that you may need, such as the one used for percent composition. For example, suppose you are told that a hydrated salt has a mass of 12.4 g. After heating to dryness, the mass of the salt is 8.5 g. Determine the percent composition of water in this salt.

$$\% \text{ composition of water} = \dfrac{12.4 \text{ g} - 8.5 \text{ g}}{12.4 \text{ g}} \times 100 = 31.5\%$$

> **Self-Check** Suppose that the percent composition of water in the salt above is actually 33%. Determine the percent error in this experiment.

Copyright © by Holt, Rinehart and Winston. All rights reserved.

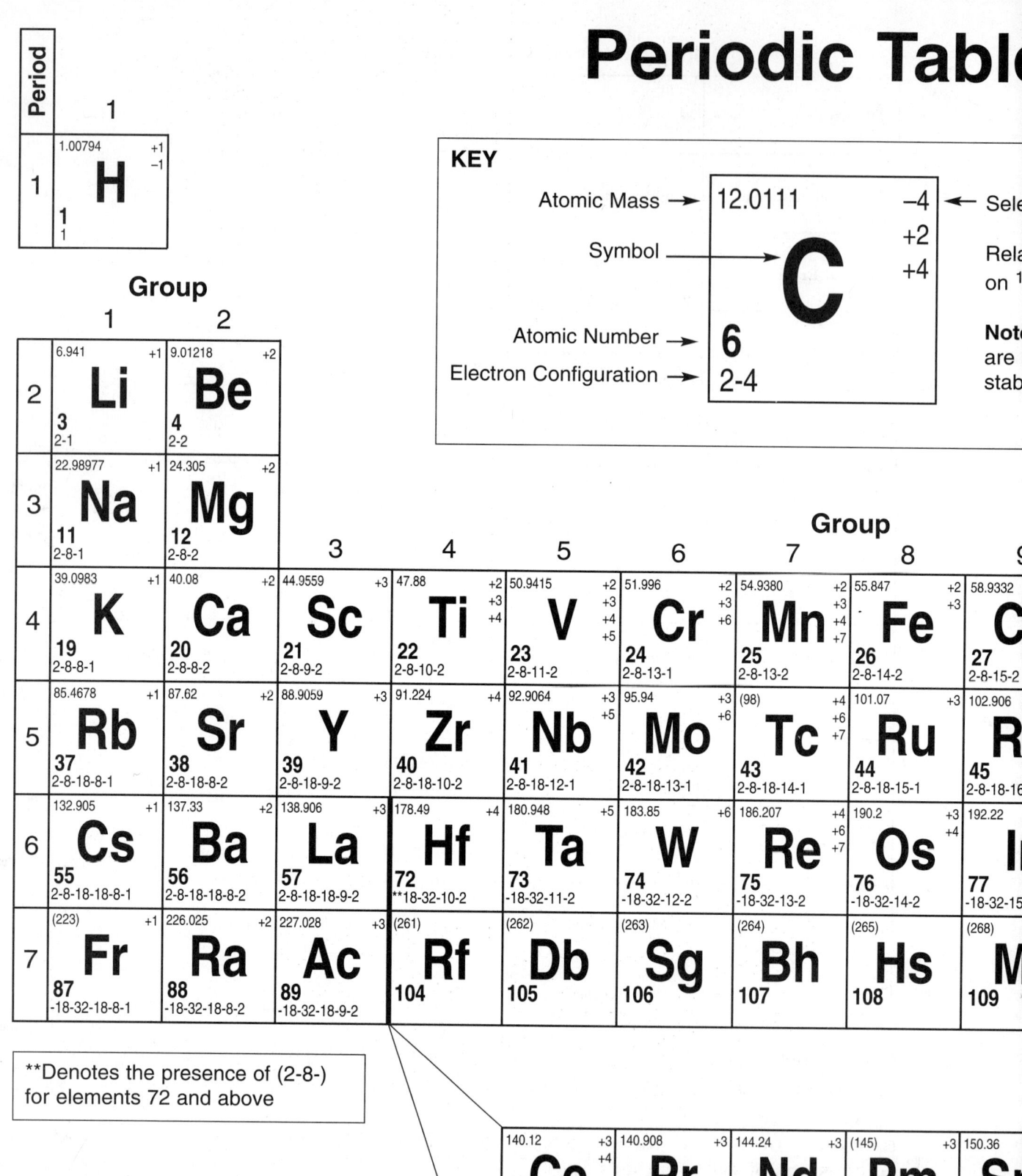

the Elements

	18
	4.00260 0
	He
	2 2

on States

masses are based

bers in parentheses
rs of the most
n isotope.

Group

13	14	15	16	17	18
10.81 +3	12.0111 −4, +2, +4	14.0067 −3, −2, −1, +1, +2, +3, +4, +5	15.9994 −2	18.998403 −1	20.179 0
B	**C**	**N**	**O**	**F**	**Ne**
5 2-3	6 2-4	7 2-5	8 2-6	9 2-7	10 2-8
26.98154 +3	28.0855 −4, +2, +4	30.97376 −3, +3, +5	32.06 −2, +4, +6	35.453 −1, +1, +3, +5, +7	39.948 0
Al	**Si**	**P**	**S**	**Cl**	**Ar**
13 2-8-3	14 2-8-4	15 2-8-5	16 2-8-6	17 2-8-7	18 2-8-8
69.72 +3	72.59 −4, +2, +4	74.9216 −3, +3, +5	78.96 −2, +4, +6	79.904 −1, +1, +5	83.80 0, +2
Ga	**Ge**	**As**	**Se**	**Br**	**Kr**
31 2-8-18-3	32 2-8-18-4	33 2-8-18-5	34 2-8-18-6	35 2-8-18-7	36 2-8-18-8
114.82 +3	118.71 +2, +4	121.75 −3, +3, +5	127.60 −2, +4, +6	126.905 −1, +1, +5, +7	131.29 0, +2, +4, +6
In	**Sn**	**Sb**	**Te**	**I**	**Xe**
49 2-8-18-18-3	50 2-8-18-18-4	51 2-8-18-18-5	52 2-8-18-18-6	53 2-8-18-18-7	54 2-8-18-18-8
204.383 +1, +3	207.2 +2, +4	208.980 +3, +5	(209) +2, +4	(210)	(222) 0
Tl	**Pb**	**Bi**	**Po**	**At**	**Rn**
81 -18-32-18-3	82 -18-32-18-4	83 -18-32-18-5	84 -18-32-18-6	85 -18-32-18-7	86 -18-32-18-8
	(285) **Uuq** 114				

Columns 11, 12 (partial left side):

11	12
63.546 +1, +2	65.39 +2
Cu	**Zn**
29 2-8-18-1	30 2-8-18-2
107.868 +1	112.41 +2
Ag	**Cd**
47 2-8-18-18-1	48 2-8-18-18-2
196.967 +1, +3	200.59 +1, +2
Au	**Hg**
79 -18-32-18-1	80 -18-32-18-2
(272) **Uuu** 111	(277) **Uub** 112

systematic names and symbols for elements of atomic numbers above 109
e used until the approval of trivial names by IUPAC.

157.25 +3	158.925 +3	162.50 +3	164.930 +3	167.26 +3	168.934 +3	173.04 +2, +3	174.967 +3
Gd	**Tb**	**Dy**	**Ho**	**Er**	**Tm**	**Yb**	**Lu**
64	65	66	67	68	69	70	71
(247) +3	(247) +3, +4	(251) +3	(252) +3	(257)	(258)	(259)	(260)
Cm	**Bk**	**Cf**	**Es**	**Fm**	**Md**	**No**	**Lr**
96	97	98	99	100	101	102	103

Reference Tables continued

Notes/Study Ideas/Answers

The information on this periodic table enables you to determine the number of protons, neutrons, and electrons in an atom of each element by using its atomic number and mass number. You can also check the electron configuration for most elements. The oxidation states are often used to determine if a redox reaction has occurred.

Remember that elements in the same group have the same number of valence electrons. Therefore, they share physical and chemical properties. Elements in the same period have the same number of occupied shells. Notice the darkened line that separates metals from nonmetals.

ANSWERS TO SELF-CHECK QUESTIONS

- 4.5 L × 315 K/273 K = 5.2 L

- 4.5 g × 334 J/g = 1503 J or 1.503 kJ
  25.5 g × 2260 J/g = 57630 J or 57.63 kJ

- 4.5 cm = 45 mm; 582 mL = 0.582 L

- part per million and molarity

- $(NH_4)_2Cr_2O_7$; $Hg_2(NO_3)_2$

- ammonium chloride (soluble); lead sulfate (insoluble); magnesium sulfide (insoluble); calcium chromate (soluble)

- This solution is unsaturated because the graph shows that about 45 g KCl can dissolve in 100 g H_2O at 60°C.

- The vapor pressure is about 58 kPa. The boiling point is 55°C.

- When this reaction occurs, 227.4 kJ of energy is absorbed. The reaction is endothermic because the ΔH value is positive.

- The reaction will not occur spontaneously because Fe is below Ca in Table J and therefore will not replace it. Ca is listed above Ni. Therefore, Ca loses electrons more readily and will serve as the anode where oxidation occurs.

Reference Tables continued

- $2HNO_3 + Ca(OH)_2 \longrightarrow Ca(NO_3)_2 + 2H_2O$
  Methyl orange would not be an appropriate indicator because it changes color in the pH range 3.2 to 4.4. The indicator must change color near pH 7 because that is the pH value produced in the neutralization reaction.

- $_{87}^{220}Fr \longrightarrow \, _{2}^{4}He + \, _{85}^{218}At$; 1.25 g will remain after four half-lives

- nonane, C_9H_{20}; butyne, C_4H_6; Butyne is unsaturated because it contains a triple bond between carbon atoms.

- halide (halocarbon)

- The bond between H and O is more polar than the bond between H and S. This can be determined by comparing the difference in their electronegativity values.
  H and O: 2.1 and 3.4 for a difference of 1.3
  H and S: 2.1 and 2.6 for a difference of 0.5

- 4.5% error

Notes/Study Ideas/Answers

The University of the State of New York

REGENTS HIGH SCHOOL EXAMINATION

PHYSICAL SETTING
CHEMISTRY

Friday, June 21, 2002 — 1:15 to 4:15 p.m., only

You are to answer *all* questions in all parts of this examination according to the directions provided in the examination booklet.

Your answer sheet for Part A and Part B–1 is the last page of this examination booklet. Turn to the last page and fold it along the perforations. Then, slowly and carefully, tear off your answer sheet and fill in the heading.

Your answer booklet for Part B–2 and Part C is stapled in the center of this examination booklet. Open the examination booklet, carefully remove your answer booklet, and close the examination booklet. Then fill in the heading of your answer booklet.

Record the number of your choice for each Part A and Part B–1 multiple-choice question on your separate answer sheet. Write your answers to the Part B–2 and Part C questions in your answer booklet. All work should be written in pen, except for graphs and drawings, which should be done in pencil. You may use scrap paper to work out the answers to the questions, but be sure to record all your answers on your answer sheet and answer booklet.

When you have completed the examination, you must sign the statement printed at the end of your separate answer sheet, indicating that you had no unlawful knowledge of the questions or answers prior to the examination and that you have neither given nor received assistance in answering any of the questions during the examination. Your answer sheet and answer booklet cannot be accepted if you fail to sign this declaration.

Notice...
A four-function or scientific calculator and a copy of the *Reference Tables for Physical Setting/Chemistry* must be available for your use while taking this examination.

DO NOT OPEN THIS EXAMINATION BOOKLET UNTIL THE SIGNAL IS GIVEN.

Part A

Answer all questions in this part.

Directions (1–30): For *each* statement or question, write on the separate answer sheet the *number* of the word or expression that, of those given, best completes the statement or answers the question. Some questions may require the use of the *Reference Tables for Physical Setting/Chemistry*.

1. What is the electron configuration of a sulfur atom in the ground state?
   (1) 2–4
   (2) 2–6
   (3) 2–8–4
   (4) 2–8–6

2. The modern model of the atom shows that electrons are
   (1) orbiting the nucleus in fixed paths
   (2) found in regions called orbitals
   (3) combined with neutrons in the nucleus
   (4) located in a solid sphere covering the nucleus

3. Which compound contains ionic bonds?
   (1) NO
   (2) NO_2
   (3) CaO
   (4) CO_2

4. All the isotopes of a given atom have
   (1) the same mass number and the same atomic number
   (2) the same mass number but different atomic numbers
   (3) different mass numbers but the same atomic number
   (4) different mass numbers and different atomic numbers

5. What are two properties of most nonmetals?
   (1) high ionization energy and poor electrical conductivity
   (2) high ionization energy and good electrical conductivity
   (3) low ionization energy and poor electrical conductivity
   (4) low ionization energy and good electrical conductivity

6. Which element is classified as a noble gas at STP?
   (1) hydrogen
   (2) oxygen
   (3) neon
   (4) nitrogen

7. The percent by mass of hydrogen in NH_3 is equal to
   (1) $\frac{17}{1} \times 100$
   (2) $\frac{17}{3} \times 100$
   (3) $\frac{1}{17} \times 100$
   (4) $\frac{3}{17} \times 100$

8. Metallic bonding occurs between atoms of
   (1) sulfur
   (2) copper
   (3) fluorine
   (4) carbon

9. Atoms of the same element that have different numbers of neutrons are classified as
   (1) charged atoms
   (2) charged nuclei
   (3) isomers
   (4) isotopes

10. Compared to the radius of a chlorine atom, the radius of a chloride ion is
    (1) larger because chlorine loses an electron
    (2) larger because chlorine gains an electron
    (3) smaller because chlorine loses an electron
    (4) smaller because chlorine gains an electron

11. Which of the following atoms has the greatest tendency to attract electrons?
    (1) barium
    (2) beryllium
    (3) boron
    (4) bromine

12. Which 5.0-milliliter sample of NH_3 will take the shape of and completely fill a closed 100.0-milliliter container?
    (1) $NH_3(s)$
    (2) $NH_3(\ell)$
    (3) $NH_3(g)$
    (4) $NH_3(aq)$

13. The strongest forces of attraction occur between molecules of
    (1) HCl
    (2) HF
    (3) HBr
    (4) HI

14 Which graph shows the pressure-temperature relationship expected for an ideal gas?

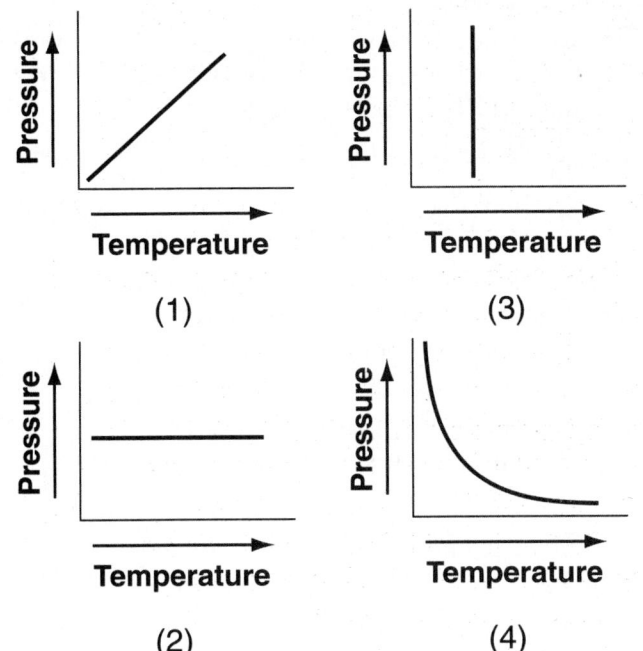

15 At the same temperature and pressure, which sample contains the same number of moles of particles as 1 liter of $O_2(g)$?
(1) 1 L Ne(g)
(2) 2 L N_2(g)
(3) 0.5 L SO_2(g)
(4) 1 L $H_2O(\ell)$

16 Which change in the temperature of a 1-gram sample of water would cause the greatest increase in the average kinetic energy of its molecules?
(1) 1°C to 10°C
(2) 10°C to 1°C
(3) 50°C to 60°C
(4) 60°C to 50°C

17 Which compound is classified as a hydrocarbon?
(1) ethane
(2) ethanol
(3) chloroethane
(4) ethanoic acid

18 Given the reaction:

$Mg(s) + 2 H^+(aq) + 2 Cl^-(aq) \rightarrow$
$Mg^{2+}(aq) + 2 Cl^-(aq) + H_2(g)$

Which species undergoes oxidation?
(1) Mg(s)
(2) H^+(aq)
(3) Cl^-(aq)
(4) H_2(g)

19 Which formula is an isomer of butane?

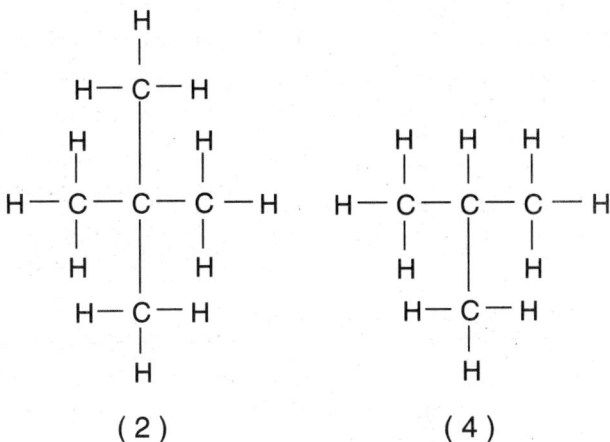

20 Which particles are gained and lost during a redox reaction?
(1) electrons
(2) protons
(3) neutrons
(4) positrons

21 What is the oxidation number of chromium in $K_2Cr_2O_7$?
(1) +12
(2) +2
(3) +3
(4) +6

22 Which process requires an external power source?
(1) neutralization
(2) synthesis
(3) fermentation
(4) electrolysis

23 A substance that conducts an electrical current when dissolved in water is called
(1) a catalyst
(2) a metalloid
(3) a nonelectrolyte
(4) an electrolyte

24 Which product of nuclear decay has mass but no charge?
(1) alpha particles
(2) neutrons
(3) gamma rays
(4) beta positrons

25 Given the reaction:

HCl(aq) + LiOH(aq) → HOH(ℓ) + LiCl(aq)

The reaction is best described as

(1) neutralization
(2) synthesis
(3) decomposition
(4) oxidation-reduction

26 Which ion is produced when an Arrhenius base is dissolved in water?

(1) H^+, as the only positive ion in solution
(2) H_3O^+, as the only positive ion in solution
(3) OH^-, as the only negative ion in solution
(4) H^-, as the only negative ion in solution

27 The change that is undergone by an atom of an element made radioactive by bombardment with high-energy protons is called

(1) natural transmutation
(2) artificial transmutation
(3) natural decay
(4) radioactive decay

Note that questions 28 through 30 have only three choices.

28 As ice melts at standard pressure, its temperature remains at 0°C until it has completely melted. Its potential energy

(1) decreases
(2) increases
(3) remains the same

29 As a sample of the radioactive isotope ^{131}I decays, its half-life

(1) decreases
(2) increases
(3) remains the same

30 As an atom becomes an ion, its mass number

(1) decreases
(2) increases
(3) remains the same

Part B–1

Answer all questions in this part.

Directions (31–50): For *each* statement or question, write on the separate answer sheet the *number* of the word or expression that, of those given, best completes the statement or answers the question. Some questions may require the use of the *Reference Tables for Physical Setting/Chemistry*.

31 In which shell are the valence electrons of the elements in Period 2 found?

(1) 1 (3) 3
(2) 2 (4) 4

32 Which of the following Group 15 elements has the greatest metallic character?

(1) nitrogen (3) antimony
(2) phosphorus (4) bismuth

33 The number of neutrons in the nucleus of an atom can be determined by

(1) adding the atomic number to the mass number
(2) subtracting the atomic number from the mass number
(3) adding the mass number to the atomic mass
(4) subtracting the mass number from the atomic number

34 A compound has a gram formula mass of 56 grams per mole. What is the molecular formula for this compound?

(1) CH_2 (3) C_3H_6
(2) C_2H_4 (4) C_4H_8

35 Given the equilibrium reaction at STP:

$$N_2O_4(g) \rightleftharpoons 2\ NO_2(g)$$

Which statement correctly describes this system?

(1) The forward and reverse reaction rates are equal.
(2) The forward and reverse reaction rates are both increasing.
(3) The concentrations of N_2O_4 and NO_2 are equal.
(4) The concentrations of N_2O_4 and NO_2 are both increasing.

36 What is the total number of oxygen atoms in the formula $MgSO_4 \cdot 7\ H_2O$? [The • represents seven units of H_2O attached to one unit of $MgSO_4$.]

(1) 11 (3) 5
(2) 7 (4) 4

37 Given the reaction:

$$6\ CO_2 + 6\ H_2O \rightarrow C_6H_{12}O_6 + 6\ O_2$$

What is the total number of moles of water needed to make 2.5 moles of $C_6H_{12}O_6$?

(1) 2.5 (3) 12
(2) 6.0 (4) 15

38 A student calculated the percent by mass of water in a hydrate as 14.2%. A hydrate is a compound that contains water as part of its crystal structure. If the accepted value is 14.7%, the student's percent error was

(1) $\frac{0.5}{14.2} \times 100$ (3) $\frac{0.5}{14.7} \times 100$
(2) $\frac{14.7}{14.2} \times 100$ (4) $\frac{14.2}{14.7} \times 100$

39 Which of the following ions has the *smallest* radius?

(1) F^- (3) K^+
(2) Cl^- (4) Ca^{2+}

40 According to Reference Table *G*, which solution is saturated at 30°C?

(1) 12 grams of $KClO_3$ in 100 grams of water
(2) 12 grams of $KClO_3$ in 200 grams of water
(3) 30 grams of NaCl in 100 grams of water
(4) 30 grams of NaCl in 200 grams of water

41 The gram formula mass of NH_4Cl is

(1) 22.4 g/mole (3) 53.5 g/mole
(2) 28.0 g/mole (4) 95.5 g/mole

42 What is the molarity of a solution that contains 0.50 mole of NaOH in 0.50 liter of solution?

(1) 1.0 M (3) 0.25 M
(2) 2.0 M (4) 0.50 M

43 Given:

● = particle X
○ = particle Y

Which diagram represents a mixture?

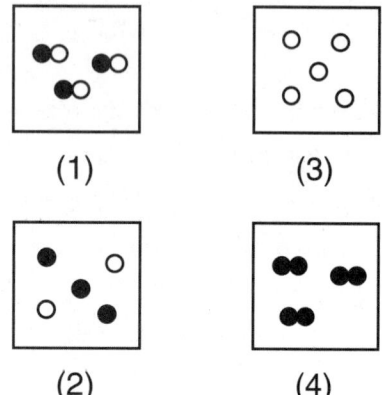

44 Which process is accompanied by a *decrease* in entropy?

(1) boiling of water
(2) condensing of water vapor
(3) subliming of iodine
(4) melting of ice

45 If 5.0 milliliters of a 0.20 M HCl solution is required to neutralize exactly 10. milliliters of NaOH, what is the concentration of the base?

(1) 0.10 M (3) 0.30 M
(2) 0.20 M (4) 0.40 M

46 Exactly how much time must elapse before 16 grams of potassium-42 decays, leaving 2 grams of the original isotope?

(1) 8 × 12.4 hours (3) 3 × 12.4 hours
(2) 2 × 12.4 hours (4) 4 × 12.4 hours

47 Which mass measurement contains four significant figures?

(1) 0.086 g (3) 1003 g
(2) 0.431 g (4) 3870 g

48 Which pair of compounds are alcohols?

(1) $H-\underset{H}{\underset{|}{C}}-\underset{OH}{\underset{|}{C}}-H$ and $H-\underset{H}{\underset{|}{C}}-\underset{H}{\underset{|}{C}}-\underset{H}{\underset{|}{C}}-OH$

(2) $H-\underset{H}{\underset{|}{C}}-\underset{OH}{\underset{|}{C}}-H$ and $H-\underset{H}{\underset{|}{C}}-C\underset{H}{\overset{O}{\diagup\!\!\!\diagdown}}$

(3) $\underset{HO}{\overset{O}{\diagdown\!\!\!\diagup}}C-C\underset{OH}{\overset{O}{\diagup\!\!\!\diagdown}}$ and $H-\underset{H}{\underset{|}{C}}-\underset{OH}{\underset{|}{C}}-\underset{H}{\underset{|}{C}}-H$

(4) $H-\underset{H}{\underset{|}{C}}-C\underset{H}{\overset{O}{\diagup\!\!\!\diagdown}}$ and $H-\underset{H}{\underset{|}{C}}-C\underset{OH}{\overset{O}{\diagup\!\!\!\diagdown}}$

49 The process of joining many small molecules into larger molecules is called

(1) neutralization (3) saponification
(2) polymerization (4) substitution

50 The diagram below represents a portion of a 100-milliliter graduated cylinder.

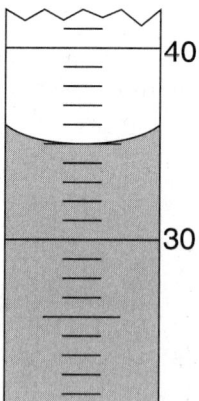

What is the reading of the meniscus?

(1) 35.0 mL (3) 44.0 mL
(2) 36.0 mL (4) 45.0 mL

Part B–2

Answer all questions in this part.

Directions (51–57): Record your answers in the spaces provided in your answer booklet. Some questions may require the use of the *Reference Tables for Physical Setting/Chemistry*.

51 In the box provided *in your answer booklet,* draw the electron-dot (Lewis) structure of an atom of calcium. [1]

52 In the box provided *in your answer booklet,* draw the electron-dot (Lewis) structure of an atom of chlorine. [1]

53 In the box provided *in your answer booklet,* draw the electron-dot (Lewis) structure of calcium chloride. [2]

54 A student is given two beakers, each containing an equal amount of clear, odorless liquid. One solution is acidic and the other is basic.

    a State *two* safe methods of distinguishing the acid solution from the base solution. [2]
    b For *each* method, state the results of both the testing of the acid solution *and* the testing of the base solution. [2]

Base your answers to questions 55 and 56 on the information and diagram below, which represent the changes in potential energy that occur during the given reaction.

Given the reaction: $A + B \rightarrow C$

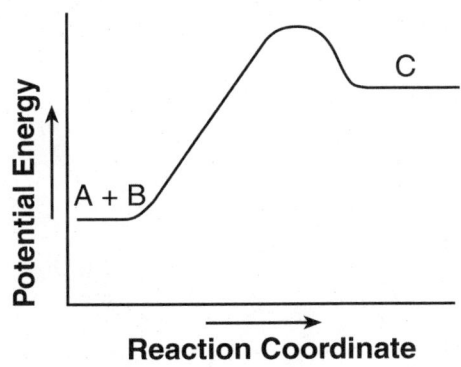

55 Does the diagram illustrate an exothermic or an endothermic reaction? State one reason, in terms of energy, to support your answer. [2]

56 On the diagram provided *in your answer booklet,* draw a dashed line to indicate a potential energy curve for the reaction if a catalyst is added. [1]

57 Given the reaction at equilibrium:

$$N_2(g) + 3\,H_2(g) \rightleftharpoons 2\,NH_3(g) + 92.05\text{ kJ}$$

    a State the effect on the number of moles of $N_2(g)$ if the temperature of the system is increased. [1]

    b State the effect on the number of moles of $H_2(g)$ if the pressure on the system is increased. [1]

    c State the effect on the number of moles of $NH_3(g)$ if a catalyst is introduced into the reaction system. Explain why this occurs. [2]

Part C

Answer all questions in this part.

Directions (58–75): Record your answers in the spaces provided in your answer booklet. Some questions may require the use of the *Reference Tables for Physical Setting/ Chemistry*.

Base your answers to questions 58 through 60 on the information below.

In the modern model of the atom, each atom is composed of three major subatomic (or fundamental) particles.

58 Name the subatomic particles contained in the nucleus of the atom.   [1]

59 State the charge associated with *each* type of subatomic particle contained in the nucleus of the atom.   [1]

60 What is the net charge of the nucleus?   [1]

Base your answers to questions 61 through 63 on the information below.

Testing of an unknown solid shows that it has the properties listed below.

(1) low melting point
(2) nearly insoluble in water
(3) nonconductor of electricity
(4) relatively soft solid

61 State the type of bonding that would be expected in the particles of this substance.   [1]

62 Explain in terms of attractions between particles why the unknown solid has a low melting point.   [1]

63 Explain why the particles of this substance are nonconductors of electricity.   [1]

Base your answers to questions 64 through 66 on the information below.

A hot pack contains chemicals that can be activated to produce heat. A cold pack contains chemicals that feel cold when activated.

64 Based on energy flow, state the type of chemical change that occurs in a hot pack. [1]

65 A cold pack is placed on an injured leg. Indicate the direction of the flow of energy between the leg and the cold pack.   [1]

66 What is the Law of Conservation of Energy? Describe how the Law of Conservation of Energy applies to the chemical reaction that occurs in the hot pack.   [2]

Base your answers to questions 67 through 69 on the table below, which shows the electronegativity of selected elements of Period 2 of the Periodic Table.

Element	Atomic Number	Electronegativity (g/mL)
Beryllium	4	1.6
Boron	5	2.0
Carbon	6	2.6
Fluorine	9	4.0
Lithium	3	1.0
Oxygen	8	3.4

67 On the grid provided *in your answer booklet,* set up a scale for electronegativity on the y-axis. Plot the data by drawing a best-fit line.   [2]

68 Using the graph, predict the electronegativity of nitrogen.   [1]

69 For these elements, state the trend in electronegativity in terms of atomic number.   [1]

Base your answers to questions 70 through 75 on the following redox reaction, which occurs spontaneously in an electrochemical cell.

$$Zn + Cr^{3+} \rightarrow Zn^{2+} + Cr$$

70 Write the half-reaction for the reduction that occurs.   [1]

71 Write the half-reaction for the oxidation that occurs.   [1]

72 *In your answer booklet,* balance the equation using the *smallest* whole-number coefficients.   [1]

73 Which species loses electrons and which species gains electrons?   [1]

74 Which half-reaction occurs at the cathode?   [1]

75 State what happens to the number of protons in a Zn atom when it changes to Zn^{2+} as the redox reaction occurs.   [1]

The University of the State of New York

REGENTS HIGH SCHOOL EXAMINATION

PHYSICAL SETTING
CHEMISTRY

Friday, June 21, 2002 — 1:15 to 4:15 p.m., only

ANSWER SHEET

Student .... Sex: ☐ Male ☐ Female Grade ....

Teacher .... School ....

Record your answers to Part A and Part B–1 on this answer sheet.

Part A			Part B–1	
1	11	21	31	41 ....
2	12	22	32	42 ....
3	13	23	33	43 ....
4	14	24	34	44 ....
5	15	25	35	45 ....
6	16	26	36	46 ....
7	17	27	37	47 ....
8	18	28	38	48 ....
9	19	29	39	49 ....
10	20	30	40	50 ....

Part A Score ☐

Part B–1 Score ☐

Write your answers to Part B–2 and Part C in your answer booklet.

The declaration below should be signed when you have completed the examination.

I do hereby affirm, at the close of this examination, that I had no unlawful knowledge of the questions or answers prior to the examination and that I have neither given nor received assistance in answering any of the questions during the examination.

_____

Signature

The University of the State of New York

REGENTS HIGH SCHOOL EXAMINATION

PHYSICAL SETTING
CHEMISTRY

Wednesday, January 29, 2003 — 9:15 a.m. to 12:15 p.m., only

You are to answer *all* questions in all parts of this examination according to the directions provided in the examination booklet.

Your answer sheet for Part A and Part B–1 is the last page of this examination booklet. Turn to the last page and fold it along the perforations. Then, slowly and carefully, tear off your answer sheet and fill in the heading.

Your answer booklet for Part B–2 and Part C is stapled in the center of this examination booklet. Open the examination booklet, carefully remove your answer booklet, and close the examination booklet. Then fill in the heading of your answer booklet.

Record the number of your choice for each Part A and Part B–1 multiple-choice question on your separate answer sheet. Write your answers to the Part B–2 and Part C questions in your answer booklet. All work should be written in pen, except for graphs and drawings, which should be done in pencil. You may use scrap paper to work out the answers to the questions, but be sure to record all your answers on your separate answer sheet and in your answer booklet.

When you have completed the examination, you must sign the statement printed at the end of your separate answer sheet, indicating that you had no unlawful knowledge of the questions or answers prior to the examination and that you have neither given nor received assistance in answering any of the questions during the examination. Your answer sheet and answer booklet cannot be accepted if you fail to sign this declaration.

Notice...

A four-function or scientific calculator and a copy of the *Reference Tables for Physical Setting/Chemistry* must be available for your use while taking this examination.

DO NOT OPEN THIS EXAMINATION BOOKLET UNTIL THE SIGNAL IS GIVEN.

Part A

Answer all questions in this part.

Directions (1–30): For *each* statement or question, write on the separate answer sheet the *number* of the word or expression that, of those given, best completes the statement or answers the question. Some questions may require the use of the *Reference Tables for Physical Setting/Chemistry*.

1 Which statement best describes electrons?
  (1) They are positive subatomic particles and are found in the nucleus.
  (2) They are positive subatomic particles and are found surrounding the nucleus.
  (3) They are negative subatomic particles and are found in the nucleus.
  (4) They are negative subatomic particles and are found surrounding the nucleus.

2 During a flame test, ions of a specific metal are heated in the flame of a gas burner. A characteristic color of light is emitted by these ions in the flame when the electrons
  (1) gain energy as they return to lower energy levels
  (2) gain energy as they move to higher energy levels
  (3) emit energy as they return to lower energy levels
  (4) emit energy as they move to higher energy levels

3 In which list are the elements arranged in order of increasing atomic mass?
  (1) Cl, K, Ar   (3) Te, I, Xe
  (2) Fe, Co, Ni   (4) Ne, F, Na

4 In which compound does chlorine have the highest oxidation number?
  (1) NaClO   (3) $NaClO_3$
  (2) $NaClO_2$ (4) $NaClO_4$

5 Which event must *always* occur for a chemical reaction to take place?
  (1) formation of a precipitate
  (2) formation of a gas
  (3) effective collisions between reacting particles
  (4) addition of a catalyst to the reaction system

6 Which Group of the Periodic Table contains atoms with a stable outer electron configuration?
  (1) 1   (3) 16
  (2) 8   (4) 18

7 From which of these atoms in the ground state can a valence electron be removed using the *least* amount of energy?
  (1) nitrogen   (3) oxygen
  (2) carbon   (4) chlorine

8 What is the percent by mass of oxygen in H_2SO_4? [formula mass = 98]
  (1) 16%   (3) 65%
  (2) 33%   (4) 98%

9 An atom of carbon-12 and an atom of carbon-14 differ in
  (1) atomic number
  (2) mass number
  (3) nuclear charge
  (4) number of electrons

10 The strength of an atom's attraction for the electrons in a chemical bond is the atom's
  (1) electronegativity   (3) heat of reaction
  (2) ionization energy   (4) heat of formation

11 Which type or types of change, if any, can reach equilibrium?
  (1) a chemical change, only
  (2) a physical change, only
  (3) both a chemical and a physical change
  (4) neither a chemical nor a physical change

12 An increase in the average kinetic energy of a sample of copper atoms occurs with an increase in
  (1) concentration   (3) pressure
  (2) temperature   (4) volume

13 The empirical formula of a compound is CH_2. Which molecular formula is correctly paired with a structural formula for this compound?

(1) C_2H_4 H—C—C—H with H, H below

(2) C_2H_4 H—C=C—H with H, H below

(3) C_3H_8 H—C—C—C—H with H, H, H above and H, H, H below

(4) C_3H_8 H—C=C—C—H with H, H, H above and H, H, H below

14 Given the equation:

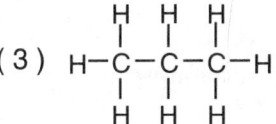

This equation represents the formation of a

(1) fluoride ion, which is smaller in radius than a fluorine atom
(2) fluoride ion, which is larger in radius than a fluorine atom
(3) fluorine atom, which is smaller in radius than a fluoride ion
(4) fluorine atom, which is larger is radius than a fluoride ion

15 The high electrical conductivity of metals is primarily due to

(1) high ionization energies
(2) filled energy levels
(3) mobile electrons
(4) high electronegativities

16 One similarity between all mixtures and compounds is that both

(1) are heterogeneous
(2) are homogeneous
(3) combine in a definite ratio
(4) consist of two or more substances

17 Which phase change results in the release of energy?

(1) $H_2O(s) \rightarrow H_2O(\ell)$
(2) $H_2O(s) \rightarrow H_2O(g)$
(3) $H_2O(\ell) \rightarrow H_2O(g)$
(4) $H_2O(g) \rightarrow H_2O(\ell)$

18 Which compound has an isomer?

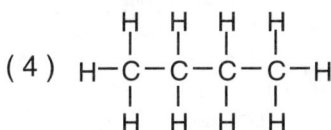

19 What occurs when NaCl(s) is added to water?

(1) The boiling point of the solution increases, and the freezing point of the solution decreases.
(2) The boiling point of the solution increases, and the freezing point of the solution increases.
(3) The boiling point of the solution decreases, and the freezing point of the solution decreases.
(4) The boiling point of the solution decreases, and the freezing point of the solution increases.

20 Which radioisotope is a beta emitter?

(1) ^{90}Sr (3) ^{37}K
(2) ^{220}Fr (4) ^{238}U

21 When a mixture of water, sand, and salt is filtered, what passes through the filter paper?

(1) water, only
(2) water and sand, only
(3) water and salt, only
(4) water, sand, and salt

22 A hydrate is a compound that includes water molecules within its crystal structure. During an experiment to determine the percent by mass of water in a hydrated crystal, a student found the mass of the hydrated crystal to be 4.10 grams. After heating to constant mass, the mass was 3.70 grams. What is the percent by mass of water in this crystal?

(1) 90.%
(2) 11%
(3) 9.8%
(4) 0.40%

23 Which of these 1 M solutions will have the highest pH?

(1) NaOH
(2) CH_3OH
(3) HCl
(4) NaCl

24 Which physical property makes it possible to separate the components of crude oil by means of distillation?

(1) melting point
(2) conductivity
(3) solubility
(4) boiling point

25 In saturated hydrocarbons, carbon atoms are bonded to each other by

(1) single covalent bonds, only
(2) double covalent bonds, only
(3) alternating single and double covalent bonds
(4) alternating double and triple covalent bonds

26 Which formula correctly represents the product of an addition reaction between ethene and chlorine?

(1) CH_2Cl_2
(2) CH_3Cl
(3) $C_2H_4Cl_2$
(4) C_2H_3Cl

27 When a neutral atom undergoes oxidation, the atom's oxidation state

(1) decreases as it gains electrons
(2) decreases as it loses electrons
(3) increases as it gains electrons
(4) increases as it loses electrons

28 Given the equation:

$$C(s) + H_2O(g) \rightarrow CO(g) + H_2(g)$$

Which species undergoes reduction?

(1) C(s)
(2) H^+
(3) C^{2+}
(4) $H_2(g)$

29 Which equation is an example of artificial transmutation?

(1) $^9_4Be + ^4_2He \rightarrow ^{12}_6C + ^1_0n$
(2) $U + 3 F_2 \rightarrow UF_6$
(3) $Mg(OH)_2 + 2 HCl \rightarrow 2 H_2O + MgCl_2$
(4) $Ca + 2 H_2O \rightarrow Ca(OH)_2 + H_2$

30 Which species can conduct an electric current?

(1) NaOH(s)
(2) $CH_3OH(aq)$
(3) $H_2O(s)$
(4) HCl(aq)

Part B–1

Answer all questions in this part.

Directions (31–50): For *each* statement or question, write on the separate answer sheet the *number* of the word or expression that, of those given, best completes the statement or answers the question. Some questions may require the use of the *Reference Tables for Physical Setting/Chemistry*.

31 According to Table *N*, which radioactive isotope is best for determining the actual age of Earth?

(1) ^{238}U (3) ^{60}Co
(2) ^{90}Sr (4) ^{14}C

32 Given the following solutions:

Solution *A*: pH of 10
Solution *B*: pH of 7
Solution *C*: pH of 5

Which list has the solutions placed in order of increasing H$^+$ concentration?

(1) *A*, *B*, *C* (3) *C*, *A*, *B*
(2) *B*, *A*, *C* (4) *C*, *B*, *A*

33 Which statement explains why nuclear waste materials may pose a problem?

(1) They frequently have short half-lives and remain radioactive for brief periods of time.
(2) They frequently have short half-lives and remain radioactive for extended periods of time.
(3) They frequently have long half-lives and remain radioactive for brief periods of time.
(4) They frequently have long half-lives and remain radioactive for extended periods of time.

34 A compound whose water solution conducts electricity and turns phenolphthalein pink is

(1) HCl (3) NaOH
(2) HC$_2$H$_3$O$_2$ (4) CH$_3$OH

35 Which of the following solids has the highest melting point?

(1) H$_2$O(s) (3) SO$_2$(s)
(2) Na$_2$O(s) (4) CO$_2$(s)

36 Hydrogen has three isotopes with mass numbers of 1, 2, and 3 and has an average atomic mass of 1.00794 amu. This information indicates that

(1) equal numbers of each isotope are present
(2) more isotopes have an atomic mass of 2 or 3 than of 1
(3) more isotopes have an atomic mass of 1 than of 2 or 3
(4) isotopes have only an atomic mass of 1

37 Which list of elements contains *two* metalloids?

(1) Si, Ge, Po, Pb (3) Si, P, S, Cl
(2) As, Bi, Br, Kr (4) Po, Sb, I, Xe

38 Given the reaction:

S(s) + O$_2$(g) → SO$_2$(g) + energy

Which diagram best represents the potential energy changes for this reaction?

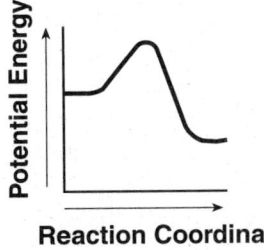
(1)

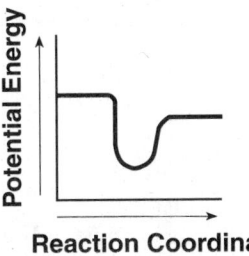
(3)

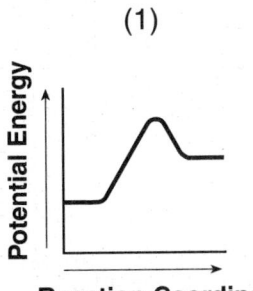
(2)

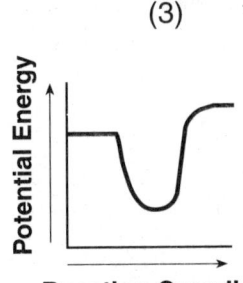
(4)

39 A chemist performs the same tests on two homogeneous white crystalline solids, A and B. The results are shown in the table below.

	Solid A	Solid B
Melting Point	High, 801°C	Low, decomposes at 186°C
Solubility in H₂O (grams per 100.0 g H₂O at 0°C)	35.7	3.2
Electrical Conductivity (in aqueous solution)	Good conductor	Nonconductor

The results of these tests suggest that
(1) both solids contain only ionic bonds
(2) both solids contain only covalent bonds
(3) solid A contains only covalent bonds and solid B contains only ionic bonds
(4) solid A contains only ionic bonds and solid B contains only covalent bonds

---

40 Solubility data for four different salts in water at 60°C are shown in the table below.

Salt	Solubility in Water at 60°C
A	10 grams / 50 grams H₂O
B	20 grams / 60 grams H₂O
C	30 grams / 120 grams H₂O
D	40 grams/ 80 grams H₂O

Which salt is most soluble at 60°C?

(1) A (3) C
(2) B (4) D

41 Which phase change represents a *decrease* in entropy?

(1) solid to liquid   (3) liquid to gas
(2) gas to liquid   (4) solid to gas

42 Given the equation:

$$2\ C_2H_2(g) + 5\ O_2(g) \rightarrow 4\ CO_2(g) + 2\ H_2O(g)$$

How many moles of oxygen are required to react completely with 1.0 mole of C_2H_2?

(1) 2.5   (3) 5.0
(2) 2.0   (4) 10

43 A student intended to make a salt solution with a concentration of 10.0 grams of solute per liter of solution. When the student's solution was analyzed, it was found to contain 8.90 grams of solute per liter of solution. What was the percent error in the concentration of the solution?

(1) 1.10%   (3) 11.0%
(2) 8.90%   (4) 18.9%

44 What is the molarity of a solution of NaOH if 2 liters of the solution contains 4 moles of NaOH?

(1) 0.5 M   (3) 8 M
(2) 2 M   (4) 80 M

45 A gas occupies a volume of 40.0 milliliters at 20°C. If the volume is increased to 80.0 milliliters at constant pressure, the resulting temperature will be equal to

(1) $20°C \times \dfrac{80.0\ \text{mL}}{40.0\ \text{mL}}$ (3) $293\ K \times \dfrac{80.0\ \text{mL}}{40.0\ \text{mL}}$

(2) $20°C \times \dfrac{40.0\ \text{mL}}{80.0\ \text{mL}}$ (4) $293\ K \times \dfrac{40.0\ \text{mL}}{80.0\ \text{mL}}$

46 According to Reference Table J, which of these metals will react most readily with 1.0 M HCl to produce $H_2(g)$?

(1) Ca   (3) Mg
(2) K   (4) Zn

47 The graph below represents the heating curve of a substance that starts as a solid below its freezing point.

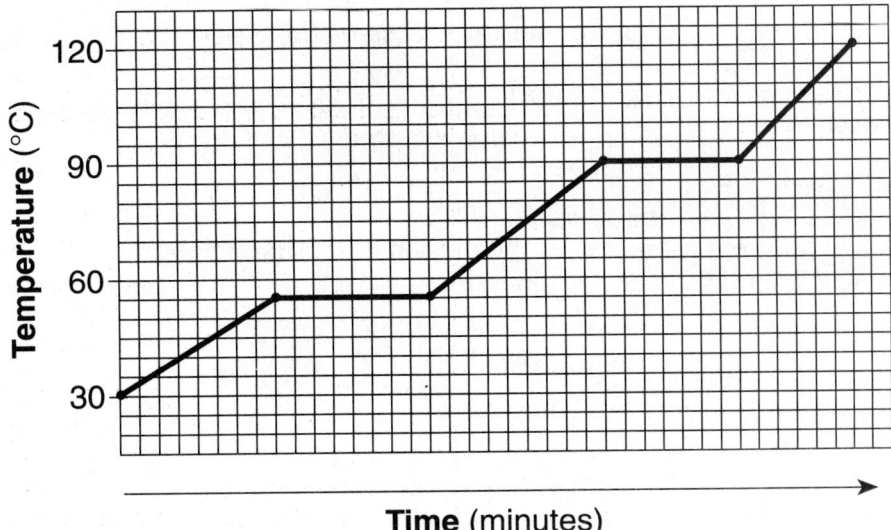

What is the melting point of this substance?
(1) 30°C
(2) 55°C
(3) 90°C
(4) 120°C

48 Given the unbalanced equation:

__Fe_2O_3 + __CO → __Fe + __CO_2

When the equation is correctly balanced using the *smallest* whole-number coefficients, what is the coefficient of CO?
(1) 1
(2) 2
(3) 3
(4) 4

49 Which type of organic compound is represented by the structural formula shown below?

```
    H   H   H
    |   |   |
H — C — C — C — H
    |   |   |
    H  OH   H
```

(1) aldehyde
(2) alcohol
(3) ether
(4) ester

50 Given the system at equilibrium:

$N_2O_4(g) + 58.1 \text{ kJ} \rightleftharpoons 2 NO_2(g)$

What will be the result of an increase in temperature at constant pressure?

(1) The equilibrium will shift to the left, and the concentration of $NO_2(g)$ will decrease.
(2) The equilibrium will shift to the left, and the concentration of $NO_2(g)$ will increase.
(3) The equilibrium will shift to the right, and the concentration of $NO_2(g)$ will decrease.
(4) The equilibrium will shift to the right, and the concentration of $NO_2(g)$ will increase.

Part B–2

Answer all questions in this part.

Directions (51–61): Record your answers in the spaces provided in your answer booklet. Some questions may require the use of the *Reference Tables for Physical Setting/Chemistry*.

51 In the boxes provided *in your answer booklet:*

  a Draw *two* different compounds, one in each box, using the representations for atoms of element X and element Z given below. [1]

  Atom of element X = ○
  Atom of element Z = ●

  b Draw a mixture of these two compounds. [1]

52 At equilibrium, nitrogen, hydrogen, and ammonia gases form a mixture in a sealed container. The data table below gives some characteristics of these substances.

Data Table

Gas	Boiling Point	Melting Point	Solubility in Water
Nitrogen	–196°C	–210°C	insoluble
Hydrogen	–252°C	–259°C	insoluble
Ammonia	–33°C	–78°C	soluble

Describe how to separate ammonia from hydrogen and nitrogen. [1]

Base your answers to questions 53 through 55 on the diagram of a voltaic cell provided *in your answer booklet* and on your knowledge of chemistry.

53 On the diagram provided *in your answer booklet,* indicate with one or more arrows the direction of electron flow through the wire. [1]

54 Write an equation for the half-reaction that occurs at the zinc electrode. [1]

55 Explain the function of the salt bridge. [1]

---

56 Given the nuclear equation:

$$^{235}_{92}U + ^{1}_{0}n \rightarrow ^{142}_{56}Ba + ^{91}_{36}Kr + 3^{1}_{0}n + \text{energy}$$

  a State the type of nuclear reaction represented by the equation. [1]
  b The sum of the masses of the products is slightly less than the sum of the masses of the reactants. Explain this loss of mass. [1]
  c This process releases greater energy than an ordinary chemical reaction does. Name another type of nuclear reaction that releases greater energy than an ordinary chemical reaction. [1]

Base your answers to questions 57 through 60 on the information below.

Each molecule listed below is formed by sharing electrons between atoms when the atoms within the molecule are bonded together.

Molecule A: Cl_2
Molecule B: CCl_4
Molecule C: NH_3

57 In the box provided *in your answer booklet*, draw the electron-dot (Lewis) structure for the NH_3 molecule. [1]

58 Explain why CCl_4 is classified as a nonpolar molecule. [1]

59 Explain why NH_3 has stronger intermolecular forces of attraction than Cl_2. [1]

60 Explain how the bonding in KCl is different from the bonding in molecules A, B, and C. [1]

61 How is the bonding between carbon atoms different in unsaturated hydrocarbons and saturated hydrocarbons? [1]

Part C

Answer all questions in this part.

Directions (62–74): Record your answers in the spaces provided in your answer booklet. Some questions may require the use of the *Reference Tables for Physical Setting/Chemistry*.

Base your answers to questions 62 through 64 on the information and diagram below.

One model of the atom states that atoms are tiny particles composed of a uniform mixture of positive and negative charges. Scientists conducted an experiment where alpha particles were aimed at a thin layer of gold atoms.

Most of the alpha particles passed directly through the gold atoms. A few alpha particles were deflected from their straight-line paths. An illustration of the experiment is shown below.

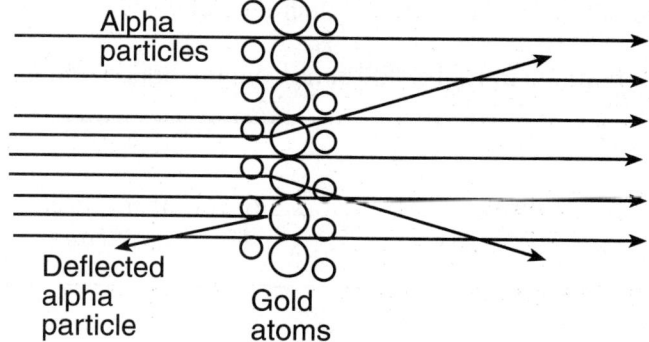

62 Most of the alpha particles passed directly through the gold atoms undisturbed. What does this evidence suggest about the structure of the gold atoms? [1]

63 A few of the alpha particles were deflected. What does this evidence suggest about the structure of the gold atoms? [1]

64 How should the original model be revised based on the results of this experiment? [1]

Base your answers to questions 65 through 67 on the information below.

When cola, a type of soda pop, is manufactured, $CO_2(g)$ is dissolved in it.

65 A capped bottle of cola contains $CO_2(g)$ under high pressure. When the cap is removed, how does pressure affect the solubility of the dissolved $CO_2(g)$? [1]

66 A glass of cold cola is left to stand 5 minutes at room temperature. How does temperature affect the solubility of the $CO_2(g)$? [1]

67 *a* In the space provided *in your answer booklet,* draw a set of axes and label one of them "Solubility" and the other "Temperature." [1]

   b Draw a line to indicate the solubility of $CO_2(g)$ versus temperature on the axes drawn in part *a*. [1]

Base your answers to questions 68 through 70 on the graph below, which shows the vapor pressure curves for liquids *A* and *B*.

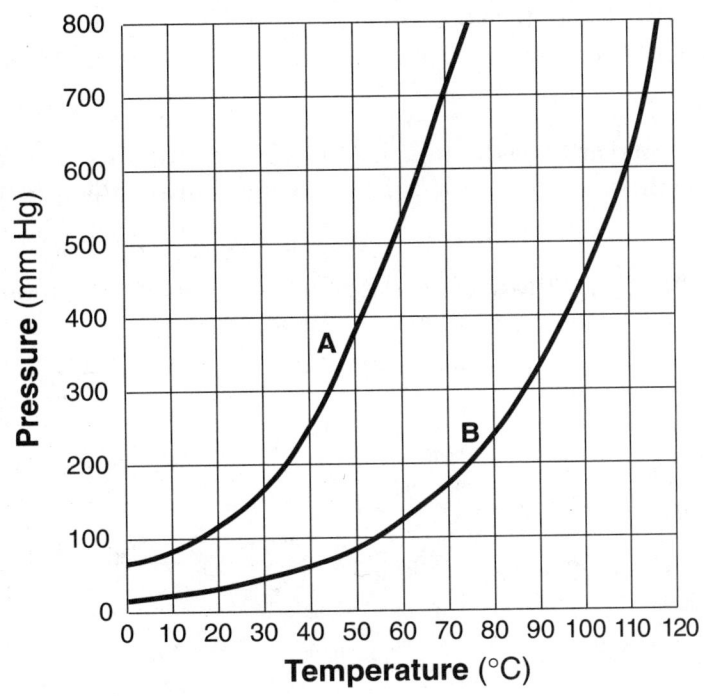

68 What is the vapor pressure of liquid *A* at 70°C? Your answer must include correct units. [2]

69 At what temperature does liquid *B* have the same vapor pressure as liquid *A* at 70°C? Your answer must include correct units. [2]

70 Which liquid will evaporate more rapidly? Explain your answer in terms of intermolecular forces. [2]

Base your answers to question 71 through 74 on the information and data table below.

A titration setup was used to determine the unknown molar concentration of a solution of NaOH. A 1.2 M HCl solution was used as the titration standard. The following data were collected.

	Trial 1	Trial 2	Trial 3	Trial 4
Amount of HCl Standard Used	10.0 mL	10.0 mL	10.0 mL	10.0 mL
Initial NaOH Buret Reading	0.0 mL	12.2 mL	23.2 mL	35.2 mL
Final NaOH Buret Reading	12.2 mL	23.2 mL	35.2 mL	47.7 mL

71 Calculate the volume of NaOH solution used to neutralize 10.0 mL of the standard HCl solution in trial 3. Show your work. [2]

72 According to Reference Table M, what indicator would be most appropriate in determining the end point of this titration? Give one reason for choosing this indicator. [2]

73 Calculate the average molarity of the unknown NaOH solution for all four trials. Your answer must include the correct number of significant figures and correct units. [3]

74 Explain why it is better to use the average data from multiple trials rather than the data from a single trial to calculate the results of the titration. [1]

The University of the State of New York

REGENTS HIGH SCHOOL EXAMINATION

PHYSICAL SETTING
CHEMISTRY

Wednesday, January 29, 2003 — 9:15 a.m. to 12:15 p.m., only

ANSWER SHEET

Student .... Sex: ☐ Male ☐ Female Grade ....

Teacher .... School ....

Record your answers to Part A and Part B–1 on this answer sheet.

Part A			Part B–1	
1	11	21	31	41 ....
2	12	22	32	42 ....
3	13	23	33	43 ....
4	14	24	34	44 ....
5	15	25	35	45 ....
6	16	26	36	46 ....
7	17	27	37	47 ....
8	18	28	38	48 ....
9	19	29	39	49 ....
10	20	30	40	50 ....

Part A Score ☐

Part B–1 Score ☐

Write your answers to Part B–2 and Part C in your answer booklet.

The declaration below should be signed when you have completed the examination.

I do hereby affirm, at the close of this examination, that I had no unlawful knowledge of the questions or answers prior to the examination and that I have neither given nor received assistance in answering any of the questions during the examination.

_____
Signature

The University of the State of New York

REGENTS HIGH SCHOOL EXAMINATION

PHYSICAL SETTING
CHEMISTRY

Tuesday, June 24, 2003 — 9:15 a.m. to 12:15 p.m., only

This is a test of your knowledge of chemistry. Use that knowledge to answer all questions in this examination. Some questions may require the use of the *Reference Tables for Physical Setting/Chemistry*. You are to answer *all* questions in all parts of this examination according to the directions provided in the examination booklet.

Your answer sheet for Part A and Part B–1 is the last page of this examination booklet. Turn to the last page and fold it along the perforations. Then, slowly and carefully, tear off your answer sheet and fill in the heading.

The answers to the questions in Part B–2 and Part C are to be written in your separate answer booklet. Be sure to fill in the heading on the front of your answer booklet.

Record the number of your choice for each Part A and Part B–1 multiple-choice question on your separate answer sheet. Write your answers to the Part B–2 and Part C questions in your answer booklet. All work should be written in pen, except for graphs and drawings, which should be done in pencil. You may use scrap paper to work out the answers to the questions, but be sure to record all your answers on your separate answer sheet and in your answer booklet.

When you have completed the examination, you must sign the statement printed at the end of your separate answer sheet, indicating that you had no unlawful knowledge of the questions or answers prior to the examination and that you have neither given nor received assistance in answering any of the questions during the examination. Your answer sheet and answer booklet cannot be accepted if you fail to sign this declaration.

Notice...

A four-function or scientific calculator and a copy of the *Reference Tables for Physical Setting/Chemistry* must be available for your use while taking this examination.

DO NOT OPEN THIS EXAMINATION BOOKLET UNTIL THE SIGNAL IS GIVEN.

Part A

Answer all questions in this part.

Directions (1–35): For *each* statement or question, write on the separate answer sheet the *number* of the word or expression that, of those given, best completes the statement or answers the question. Some questions may require the use of the *Reference Tables for Physical Setting/Chemistry*.

1 The atomic number of an atom is always equal to the number of its

(1) protons, only
(2) neutrons, only
(3) protons plus neutrons
(4) protons plus electrons

2 Which subatomic particle has no charge?

(1) alpha particle
(2) beta particle
(3) neutron
(4) electron

3 When the electrons of an excited atom return to a lower energy state, the energy emitted can result in the production of

(1) alpha particles
(2) isotopes
(3) protons
(4) spectra

4 The atomic mass of an element is calculated using the

(1) atomic number and the ratios of its naturally occurring isotopes
(2) atomic number and the half-lives of each of its isotopes
(3) masses and the ratios of its naturally occurring isotopes
(4) masses and the half-lives of each of its isotopes

5 The region that is the most probable location of an electron in an atom is

(1) the nucleus
(2) an orbital
(3) the excited state
(4) an ion

6 Which is a property of most nonmetallic solids?

(1) high thermal conductivity
(2) high electrical conductivity
(3) brittleness
(4) malleability

7 Alpha particles are emitted during the radioactive decay of

(1) carbon-14
(2) neon-19
(3) calcium-37
(4) radon-222

8 Which is an empirical formula?

(1) P_2O_5
(2) P_4O_6
(3) C_2H_4
(4) C_3H_6

9 Which substance can be decomposed by a chemical change?

(1) Co
(2) CO
(3) Cr
(4) Cu

10 The percent by mass of calcium in the compound calcium sulfate ($CaSO_4$) is approximately

(1) 15%
(2) 29%
(3) 34%
(4) 47%

11 What is represented by the dots in a Lewis electron-dot diagram of an atom of an element in Period 2 of the Periodic Table?

(1) the number of neutrons in the atom
(2) the number of protons in the atom
(3) the number of valence electrons in the atom
(4) the total number of electrons in the atom

12 Which type of chemical bond is formed between two atoms of bromine?

(1) metallic
(2) hydrogen
(3) ionic
(4) covalent

13 Which of these formulas contains the most polar bond?

(1) H–Br
(2) H–Cl
(3) H–F
(4) H–I

14 According to Table *F*, which of these salts is *least* soluble in water?
   (1) LiCl
   (2) RbCl
   (3) $FeCl_2$
   (4) $PbCl_2$

15 Which of these terms refers to matter that could be heterogeneous?
   (1) element
   (2) mixture
   (3) compound
   (4) solution

16 In which material are the particles arranged in a regular geometric pattern?
   (1) $CO_2(g)$
   (2) NaCl(aq)
   (3) $H_2O(\ell)$
   (4) $C_{12}H_{22}O_{11}(s)$

17 Which change is exothermic?
   (1) freezing of water
   (2) melting of iron
   (3) vaporization of ethanol
   (4) sublimation of iodine

18 Which type of change must occur to form a compound?
   (1) chemical
   (2) physical
   (3) nuclear
   (4) phase

19 Which formula correctly represents the composition of iron (III) oxide?
   (1) FeO_3
   (2) Fe_2O_3
   (3) Fe_3O
   (4) Fe_3O_2

20 Given the reaction:

   $PbCl_2(aq) + Na_2CrO_4(aq) \rightarrow$

   $PbCrO_4(s) + 2\ NaCl(aq)$

   What is the total number of moles of NaCl formed when 2 moles of Na_2CrO_4 react completely?
   (1) 1 mole
   (2) 2 moles
   (3) 3 moles
   (4) 4 moles

21 Which hydrocarbon is saturated?
   (1) propene
   (2) ethyne
   (3) butene
   (4) heptane

22 Which statement correctly describes an endothermic chemical reaction?
   (1) The products have higher potential energy than the reactants, and the ∆H is negative.
   (2) The products have higher potential energy than the reactants, and the ∆H is positive.
   (3) The products have lower potential energy than the reactants, and the ∆H is negative.
   (4) The products have lower potential energy than the reactants, and the ∆H is positive.

23 At standard pressure when NaCl is added to water, the solution will have a
   (1) higher freezing point and a lower boiling point than water
   (2) higher freezing point and a higher boiling point than water
   (3) lower freezing point and a higher boiling point than water
   (4) lower freezing point and a lower boiling point than water

24 Which element has atoms that can form single, double, and triple covalent bonds with other atoms of the same element?
   (1) hydrogen
   (2) oxygen
   (3) fluorine
   (4) carbon

25 Which compound is an isomer of pentane?
   (1) butane
   (2) propane
   (3) methyl butane
   (4) methyl propane

26 In which substance does chlorine have an oxidation number of +1?
   (1) Cl_2
   (2) HCl
   (3) HClO
   (4) $HClO_2$

27 Which statement is true for any electrochemical cell?
   (1) Oxidation occurs at the anode, only.
   (2) Reduction occurs at the anode, only.
   (3) Oxidation occurs at both the anode and the cathode.
   (4) Reduction occurs at both the anode and the cathode.

28 Given the equation:
$$2\ Al + 3\ Cu^{2+} \rightarrow 2\ Al^{3+} + 3\ Cu$$
The reduction half-reaction is
(1) $Al \rightarrow Al^{3+} + 3e^-$
(2) $Cu^{2+} + 2e^- \rightarrow Cu$
(3) $Al + 3e^- \rightarrow Al^{3+}$
(4) $Cu^{2+} \rightarrow Cu + 2e^-$

29 Which 0.1 M solution contains an electrolyte?
(1) $C_6H_{12}O_6(aq)$
(2) $CH_3COOH(aq)$
(3) $CH_3OH(aq)$
(4) $CH_3OCH_3(aq)$

30 Which equation represents a neutralization reaction?
(1) $Na_2CO_3 + CaCl_2 \rightarrow 2\ NaCl + CaCO_3$
(2) $Ni(NO_3)_2 + H_2S \rightarrow NiS + 2\ HNO_3$
(3) $NaCl + AgNO_3 \rightarrow AgCl + NaNO_3$
(4) $H_2SO_4 + Mg(OH)_2 \rightarrow MgSO_4 + 2\ H_2O$

31 An Arrhenius acid has
(1) only hydroxide ions in solution
(2) only hydrogen ions in solution
(3) hydrogen ions as the only positive ions in solution
(4) hydrogen ions as the only negative ions in solution

32 Which type of radioactive emission has a positive charge and weak penetrating power?
(1) alpha particle
(2) beta particle
(3) gamma ray
(4) neutron

33 Which substance contains metallic bonds?
(1) $Hg(\ell)$
(2) $H_2O(\ell)$
(3) $NaCl(s)$
(4) $C_6H_{12}O_6(s)$

34 What is the name of the process in which the nucleus of an atom of one element is changed into the nucleus of an atom of a different element?
(1) decomposition
(2) transmutation
(3) substitution
(4) reduction

Note that question 35 has only three choices.

35 A catalyst is added to a system at equilibrium. If the temperature remains constant, the activation energy of the forward reaction
(1) decreases
(2) increases
(3) remains the same

Part B–1

Answer all questions in this part.

Directions (36–50): For *each* statement or question, write on the separate answer sheet the *number* of the word or expression that, of those given, best completes the statement or answers the question. Some questions may require the use of the *Reference Tables for Physical Setting/Chemistry*.

36 The nucleus of an atom of K-42 contains

(1) 19 protons and 23 neutrons
(2) 19 protons and 42 neutrons
(3) 20 protons and 19 neutrons
(4) 23 protons and 19 neutrons

37 What is the total number of electrons in a Cu$^+$ ion?

(1) 28 (3) 30
(2) 29 (4) 36

38 Which list of elements is arranged in order of increasing atomic radii?

(1) Li, Be, B, C (3) Sc, Ti, V, Cr
(2) Sr, Ca, Mg, Be (4) F, Cl, Br, I

39 Which isotope is most commonly used in the radioactive dating of the remains of organic materials?

(1) ^{14}C (3) ^{32}P
(2) ^{16}N (4) ^{37}K

40 According to Reference Table *H*, what is the vapor pressure of propanone at 45°C?

(1) 22 kPa (3) 70. kPa
(2) 33 kPa (4) 98 kPa

41 The freezing point of bromine is

(1) 539°C (3) 7°C
(2) –539°C (4) –7°C

42 Hexane (C$_6$H$_{14}$) and water do *not* form a solution. Which statement explains this phenomenon?

(1) Hexane is polar and water is nonpolar.
(2) Hexane is ionic and water is polar.
(3) Hexane is nonpolar and water is polar.
(4) Hexane is nonpolar and water is ionic.

43 The potential energy diagram below represents a reaction.

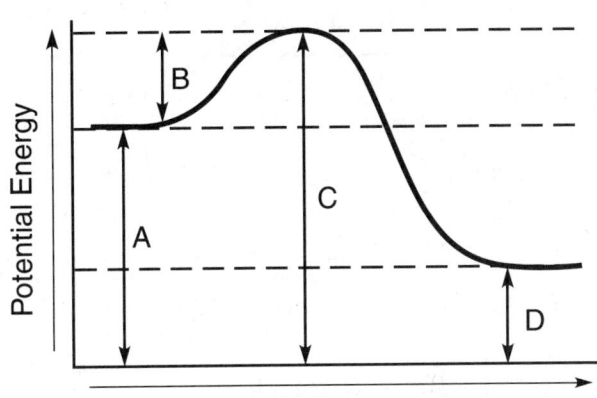

Which arrow represents the activation energy of the forward reaction?

(1) A (3) C
(2) B (4) D

44 Given the formulas of four organic compounds:

(a) 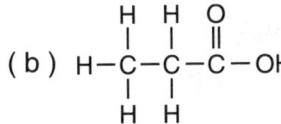   (c)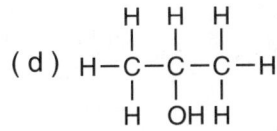

(b)
(d)

Which pair below contains an alcohol and an acid?

(1) *a* and *b* (3) *b* and *d*
(2) *a* and *c* (4) *c* and *d*

45 Which type of reaction is represented by the equation below?

Note: n and n are very large numbers equal to about 2000.

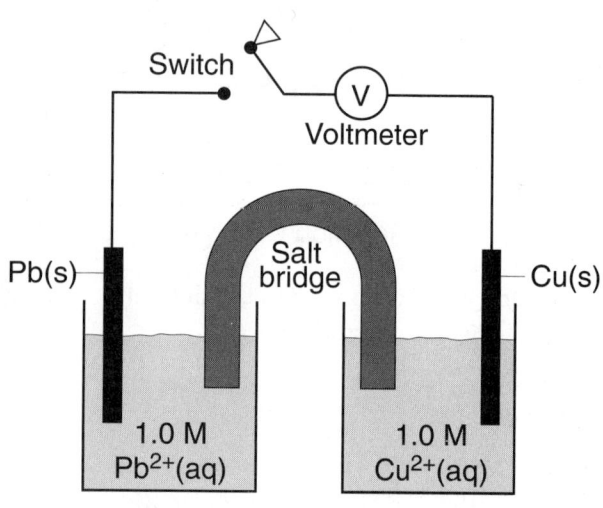

(1) esterification
(2) fermentation
(3) saponification
(4) polymerization

---

46 A diagram of a chemical cell and an equation are shown below.

Pb(s) + Cu²⁺(aq) → Pb²⁺(aq) + Cu(s)

When the switch is closed, electrons will flow from

(1) the Pb(s) to the Cu(s)
(2) the Cu(s) to the Pb(s)
(3) the Pb²⁺(aq) to the Pb(s)
(4) the Cu²⁺(aq) to the Cu(s)

47 Which ion has the same electron configuration as an atom of He?

(1) H⁻
(2) O²⁻
(3) Na⁺
(4) Ca²⁺

48 A student was given four unknown solutions. Each solution was checked for conductivity and tested with phenolphthalein. The results are shown in the data table below.

Solution	Conductivity	Color with Phenolphthalein
A	Good	Colorless
B	Poor	Colorless
C	Good	Pink
D	Poor	Pink

Based on the data table, which unknown solution could be 0.1 M NaOH?

(1) A
(2) B
(3) C
(4) D

49 In the reaction $^{239}_{93}\text{Np} \rightarrow \ ^{239}_{94}\text{Pu} + X$, what does X represent?

(1) a neutron
(2) a proton
(3) an alpha particle
(4) a beta particle

Note that question 50 has only three choices.

50 As carbon dioxide sublimes, its entropy

(1) decreases
(2) increases
(3) remains the same

Part B–2

Answer all questions in this part.

Directions (51–63): Record your answers in the spaces provided in your answer booklet. Some questions may require the use of the *Reference Tables for Physical Setting/Chemistry*.

Base your answers to questions 51 and 52 on the electron configuration table shown below.

Element	Electron Configuration
X	2–8–8–2
Y	2–8–7–3
Z	2–8–8

51 What is the total number of valence electrons in an atom of electron configuration *X*?   [1]

52 Which electron configuration represents the excited state of a calcium atom?   [1]

Base your answers to questions 53 and 54 on the information below.

Given: Samples of Na, Ar, As, Rb

53 Which *two* of the given elements have the most similar chemical properties?   [1]

54 Explain your answer in terms of the Periodic Table of the Elements.   [1]

Base your answers to questions 55 and 56 on the information below.

Diethyl ether is widely used as a solvent.

55 In the space provided *in your answer booklet*, draw the structural formula for diethyl ether.   [1]

56 In the space provided *in your answer booklet*, draw the structural formula for an alcohol that is an isomer of diethyl ether.   [1]

Base your answers to questions 57 and 58 on the information below.

Two chemistry students each combine a different metal with hydrochloric acid. Student A uses zinc, and hydrogen gas is readily produced. Student B uses copper, and no hydrogen gas is produced.

57 State one chemical reason for the different results of students A and B. [1]

58 Using Reference Table J, identify another metal that will react with hydrochloric acid to yield hydrogen gas. [1]

---

59 Given the reaction between two different elements in the gaseous state:

Box A below represents a mixture of the two reactants before the reaction occurs. The product of this reaction is a gas. In Box B provided *in your answer booklet,* draw the system after the reaction has gone to completion, based on the Law of Conservation of Matter. [2]

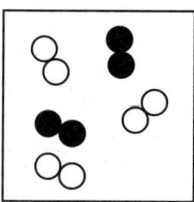

Box A

System Before Reaction

60 As a neutral sulfur atom gains two electrons, what happens to the radius of the atom? [1]

61 After a neutral sulfur atom gains two electrons, what is the resulting charge of the ion? [1]

62 *a* In the space provided *in your answer booklet,* calculate the heat released when 25.0 grams of water freezes at 0°C. Show all work. [1]
   b Record your answer with an appropriate unit. [1]

63 State one difference between voltaic cells and electrolytic cells. Include information about *both* types of cells in your answer. [1]

Part C

Answer all questions in this part.

Directions (64–79): Record your answers in the spaces provided in your answer booklet. Some questions may require the use of the *Reference Tables for Physical Setting/Chemistry*.

Base your answers to questions 64 and 65 on the diagram below, which shows a piston confining a gas in a cylinder.

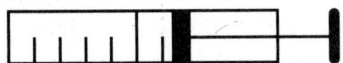

64 Using the set of axes provided *in your answer booklet,* sketch the general relationship between the pressure and the volume of an ideal gas at constant temperature. [1]

65 The gas volume in the cylinder is 6.2 milliliters and its pressure is 1.4 atmospheres. The piston is then pushed in until the gas volume is 3.1 milliliters while the temperature remains constant.

a In the space provided *in your answer booklet,* calculate the pressure, in atmospheres, after the change in volume. Show all work. [1]

b Record your answer. [1]

---

66 A student recorded the following buret readings during a titration of a base with an acid:

	Standard 0.100 M HCl	Unknown KOH
Initial reading	9.08 mL	0.55 mL
Final reading	19.09 mL	5.56 mL

a In the space provided *in your answer booklet,* calculate the molarity of the KOH. Show all work. [1]

b Record your answer to the correct number of significant figures. [1]

67 John Dalton was an English scientist who proposed that atoms were hard, indivisible spheres. In the modern model, the atom has a different internal structure.

a Identify one experiment that led scientists to develop the modern model of the atom. [1]

b Describe this experiment. [1]

c State one conclusion about the internal structure of the atom, based on this experiment. [1]

Base your answers to questions 68 through 73 on the information below and on your knowledge of chemistry.

Nuclear Waste Storage Plan for Yucca Mountain

In 1978, the U.S. Department of Energy began a study of Yucca Mountain which is located 90 miles from Las Vegas, Nevada. The study was to determine if Yucca Mountain would be suitable for a long-term burial site for high-level radioactive waste. A three-dimensional (3-D) computer scale model of the site was used to simulate the Yucca Mountain area. The computer model study for Yucca Mountain included such variables as: the possibility of earthquakes, predicted water flow through the mountain, increased rainfall due to climate changes, radioactive leakage from the waste containers, and increased temperatures from the buried waste within the containers.

The containers that will be used to store the radioactive waste are designed to last 10,000 years. Within the 10,000-year time period, cesium and strontium, the most powerful radioactive emitters, would have decayed. Other isotopes found in the waste would decay more slowly, but are not powerful radioactive emitters.

In 1998, scientists discovered that the compressed volcanic ash making up Yucca Mountain was full of cracks. Because of the arid climate, scientists assumed that rainwater would move through the cracks at a slow rate. However, when radioactive chlorine-36 was found in rock samples at levels halfway through the mountain, it was clear that rainwater had moved quickly down through Yucca Mountain. It was only 50 years earlier when this chlorine-36 isotope had contaminated rainwater during atmospheric testing of the atom bomb.

Some opponents of the Yucca Mountain plan believe that the uncertainties related to the many variables of the computer model result in limited reliability of its predictions. However, advocates of the plan believe it is safer to replace the numerous existing radioactive burial sites around the United States with the one site at Yucca Mountain. Other opponents of the plan believe that transporting the radioactive waste to Yucca Mountain from the existing 131 burial sites creates too much danger to the United States. In 2002, after years of political debate, a final legislative vote approved the development of Yucca Mountain to replace the existing 131 burial sites.

68 State one uncertainty in the computer model that limits the reliability of this computer model. [1]

69 Scientists assume that a manufacturing defect would cause at least one of the waste containers stored in the Yucca Mountain repository to leak within the first 1,000 years. State one possible effect such a leak could have on the environment near Yucca Mountain. [1]

70 State one risk associated with leaving radioactive waste in the 131 sites around the country where it is presently stored. [1]

71 If a sample of cesium-137 is stored in a waste container in Yucca Mountain, how much time must elapse until only $\frac{1}{32}$ of the original sample remains unchanged? [1]

72 The information states "Within the 10,000-year time period, cesium and strontium, the most powerful radioactive emitters, would have decayed." Use information from Reference Table N to support this statement. [1]

73 Why is water flow a crucial factor in deciding whether Yucca Mountain is a suitable burial site? [1]

Base your answers to questions 74 through 76 on the information below.

A student wishes to investigate how the reaction rate changes with a change in concentration of HCl(aq).

Given the reaction: $Zn(s) + HCl(aq) \rightarrow H_2(g) + ZnCl_2(aq)$

74 Identify the independent variable in this investigation. [1]

75 Identify one other variable that might affect the rate and should be held constant during this investigation. [1]

76 Describe the effect of increasing the concentration of HCl(aq) on the reaction rate and justify your response in terms of *collision theory*. [1]

Base your answers to questions 77 through 79 on the information below.

A truck carrying concentrated nitric acid overturns and spills its contents. The acid drains into a nearby pond. The pH of the pond water was 8.0 before the spill. After the spill, the pond water is 1,000 times more acidic.

77 Name an ion in the pond water that has increased in concentration due to this spill. [1]

78 What is the new pH of the pond water after the spill? [1]

79 What color would bromthymol blue be at this new pH? [1]

The University of the State of New York

REGENTS HIGH SCHOOL EXAMINATION

PHYSICAL SETTING
CHEMISTRY

Tuesday, June 24, 2003 — 9:15 a.m. to 12:15 p.m., only

ANSWER SHEET

Student .... Sex: ☐ Male ☐ Female Grade ....

Teacher .... School ....

Record your answers to Part A and Part B–1 on this answer sheet.

Part A			Part B–1	
1	13	25	36	44 ....
2	14	26	37	45 ....
3	15	27	38	46 ....
4	16	28	39	47 ....
5	17	29	40	48 ....
6	18	30	41	49 ....
7	19	31	42	50 ....
8	20	32	43	
9	21	33		
10	22	34		
11	23	35		
12	24			

Part A Score ☐

Part B–1 Score ☐

Write your answers to Part B–2 and Part C in your answer booklet.

The declaration below should be signed when you have completed the examination.

I do hereby affirm, at the close of this examination, that I had no unlawful knowledge of the questions or answers prior to the examination and that I have neither given nor received assistance in answering any of the questions during the examination.

Signature

The University of the State of New York

REGENTS HIGH SCHOOL EXAMINATION

PHYSICAL SETTING
CHEMISTRY

Wednesday, January 28, 2004 — 9:15 a.m. to 12:15 p.m., only

This is a test of your knowledge of chemistry. Use that knowledge to answer all questions in this examination. Some questions may require the use of the *Reference Tables for Physical Setting/Chemistry*. You are to answer *all* questions in all parts of this examination according to the directions provided in the examination booklet.

Your answer sheet for Part A and Part B–1 is the last page of this examination booklet. Turn to the last page and fold it along the perforations. Then, slowly and carefully, tear off your answer sheet and fill in the heading.

The answers to the questions in Part B–2 and Part C are to be written in your separate answer booklet. Be sure to fill in the heading on the front of your answer booklet.

Record the number of your choice for each Part A and Part B–1 multiple-choice question on your separate answer sheet. Write your answers to the Part B–2 and Part C questions in your answer booklet. All work should be written in pen, except for graphs and drawings, which should be done in pencil. You may use scrap paper to work out the answers to the questions, but be sure to record all your answers on your separate answer sheet and in your answer booklet.

When you have completed the examination, you must sign the statement printed at the end of your separate answer sheet, indicating that you had no unlawful knowledge of the questions or answers prior to the examination and that you have neither given nor received assistance in answering any of the questions during the examination. Your answer sheet and answer booklet cannot be accepted if you fail to sign this declaration.

Notice...

A four-function or scientific calculator and a copy of the *Reference Tables for Physical Setting/Chemistry* must be available for your use while taking this examination.

DO NOT OPEN THIS EXAMINATION BOOKLET UNTIL THE SIGNAL IS GIVEN.

Part A

Answer all questions in this part.

Directions (1–31): For *each* statement or question, write on the separate answer sheet the *number* of the word or expression that, of those given, best completes the statement or answers the question. Some questions may require the use of the *Reference Tables for Physical Setting/Chemistry*.

1 A neutral atom contains 12 neutrons and 11 electrons. The number of protons in this atom is

(1) 1
(2) 11
(3) 12
(4) 23

2 Isotopes of an element must have different

(1) atomic numbers
(2) mass numbers
(3) numbers of protons
(4) numbers of electrons

3 Which element is a noble gas?

(1) krypton
(2) chlorine
(3) antimony
(4) manganese

4 On the present Periodic Table of the Elements, the elements are arranged according to increasing

(1) number of oxidation states
(2) number of neutrons
(3) atomic mass
(4) atomic number

5 What is a property of most metals?

(1) They tend to gain electrons easily when bonding.
(2) They tend to lose electrons easily when bonding.
(3) They are poor conductors of heat.
(4) They are poor conductors of electricity.

6 What is the correct formula for iron (III) phosphate?

(1) FeP
(2) Fe_3P_2
(3) $FePO_4$
(4) $Fe_3(PO_4)_2$

7 The bond between Br atoms in a Br_2 molecule is

(1) ionic and is formed by the sharing of two valence electrons
(2) ionic and is formed by the transfer of two valence electrons
(3) covalent and is formed by the sharing of two valence electrons
(4) covalent and is formed by the transfer of two valence electrons

8 The amount of energy required to remove the outermost electron from a gaseous atom in the ground state is known as

(1) first ionization energy
(2) activation energy
(3) conductivity
(4) electronegativity

9 What occurs when an atom of chlorine and an atom of hydrogen become a molecule of hydrogen chloride?

(1) A chemical bond is broken and energy is released.
(2) A chemical bond is broken and energy is absorbed.
(3) A chemical bond is formed and energy is released.
(4) A chemical bond is formed and energy is absorbed.

10 Which molecule is nonpolar?

(1) H_2O
(2) NH_3
(3) CO
(4) CO_2

11 Which must be a mixture of substances?

(1) solid
(2) liquid
(3) gas
(4) solution

12 A bottle of rubbing alcohol contains both 2-propanol and water. These liquids can be separated by the process of distillation because the 2-propanol and water
(1) have combined chemically and retain their different boiling points
(2) have combined chemically and have the same boiling point
(3) have combined physically and retain their different boiling points
(4) have combined physically and have the same boiling point

13 Compared to pure water, an aqueous solution of calcium chloride has a
(1) higher boiling point and higher freezing point
(2) higher boiling point and lower freezing point
(3) lower boiling point and higher freezing point
(4) lower boiling point and lower freezing point

14 Under which conditions does a real gas behave most like an ideal gas?
(1) at low temperatures and high pressures
(2) at low temperatures and low pressures
(3) at high temperatures and high pressures
(4) at high temperatures and low pressures

15 What is the IUPAC name of the compound with the following structural formula?

$$H-\underset{\underset{H}{|}}{\overset{\overset{H}{|}}{C}}-\overset{\overset{O}{\|}}{C}-\underset{\underset{H}{|}}{\overset{\overset{H}{|}}{C}}-\underset{\underset{H}{|}}{\overset{\overset{H}{|}}{C}}-H$$

(1) propanone
(2) propanal
(3) butanone
(4) butanal

16 Which statement best explains the role of a catalyst in a chemical reaction?
(1) A catalyst is added as an additional reactant and is consumed but not regenerated.
(2) A catalyst limits the amount of reactants used.
(3) A catalyst changes the kinds of products produced.
(4) A catalyst provides an alternate reaction pathway that requires less activation energy.

17 Given the reaction at equilibrium:

$$H_2(g) + Br_2(g) \rightleftharpoons 2\ HBr(g)$$

The rate of the forward reaction is
(1) greater than the rate of the reverse reaction
(2) less than the rate of the reverse reaction
(3) equal to the rate of the reverse reaction
(4) independent of the rate of the reverse reaction

18 Which statement best explains why most atomic masses on the Periodic Table are decimal numbers?
(1) Atomic masses are determined relative to an H–1 standard.
(2) Atomic masses are determined relative to an O–16 standard.
(3) Atomic masses are a weighted average of the naturally occurring isotopes.
(4) Atomic masses are an estimated average of the artificially produced isotopes.

19 All organic compounds must contain the element
(1) phosphorus
(2) oxygen
(3) carbon
(4) nitrogen

20 Which of the following compounds has the highest boiling point?
(1) H_2O
(2) H_2S
(3) H_2Se
(4) H_2Te

21 The functional group —COOH is found in
(1) esters
(2) aldehydes
(3) alcohols
(4) organic acids

22 Which of these elements is the best conductor of electricity?
(1) S
(2) N
(3) Br
(4) Ni

23 Given the reaction:

$$2\,Al(s) + Fe_2O_3(s) \xrightarrow{heat} Al_2O_3(s) + 2\,Fe(s)$$

Which species undergoes reduction?

(1) Al
(2) Fe
(3) Al^{3+}
(4) Fe^{3+}

24 Which energy transformation occurs when an electrolytic cell is in operation?

(1) chemical energy → electrical energy
(2) electrical energy → chemical energy
(3) light energy → heat energy
(4) light energy → chemical energy

25 Which of these pH numbers indicates the highest level of acidity?

(1) 5
(2) 8
(3) 10
(4) 12

26 According to the Arrhenius theory, when a base dissolves in water it produces

(1) CO_3^{2-} as the only negative ion in solution
(2) OH^- as the only negative ion in solution
(3) NH_4^+ as the only positive ion in solution
(4) H^+ as the only positive ion in solution

27 Which compound is an electrolyte?

(1) $C_6H_{12}O_6$
(2) CH_3OH
(3) $CaCl_2$
(4) CCl_4

28 Which equation represents a spontaneous nuclear decay?

(1) $C + O_2 \to CO_2$
(2) $H_2CO_3 \to CO_2 + H_2O$
(3) $^{27}_{13}Al + ^{4}_{2}He \to ^{30}_{15}P + ^{1}_{0}n$
(4) $^{90}_{38}Sr \to ^{0}_{-1}e + ^{90}_{39}Y$

29 The stability of an isotope is based on its

(1) number of neutrons, only
(2) number of protons, only
(3) ratio of neutrons to protons
(4) ratio of electrons to protons

Note that questions 30 and 31 have only three choices.

30 As the temperature of a substance *decreases*, the average kinetic energy of its particles

(1) decreases
(2) increases
(3) remains the same

31 When an atom of phosphorus becomes a phosphide ion (P^{3-}), the radius

(1) decreases
(2) increases
(3) remains the same

Part B–1

Answer all questions in this part.

Directions (32–50): For *each* statement or question, write on the separate answer sheet the *number* of the word or expression that, of those given, best completes the statement or answers the question. Some questions may require the use of the *Reference Tables for Physical Setting/Chemistry*.

32 The data table below represents the properties determined by the analysis of substances *A*, *B*, *C*, and *D*.

Substance	Melting Point (°C)	Boiling Point (°C)	Conductivity
A	–80	–20	none
B	20	190	none
C	320	770	as solid
D	800	1250	in solution

Which substance is an ionic compound?

(1) *A*
(2) *B*
(3) *C*
(4) *D*

33 What is the total number of electrons in a Cr^{3+} ion?

(1) 18   (3) 24
(2) 21   (4) 27

34 As the atoms of the Group 17 elements in the ground state are considered from top to bottom, each successive element has

(1) the same number of valence electrons and similar chemical properties
(2) the same number of valence electrons and identical chemical properties
(3) an increasing number of valence electrons and similar chemical properties
(4) an increasing number of valence electrons and identical chemical properties

35 Which solution when mixed with a drop of bromthymol blue will cause the indicator to change from blue to yellow?

(1) 0.1 M HCl   (3) 0.1 M CH_3OH
(2) 0.1 M NH_3 (4) 0.1 M NaOH

36 What is the empirical formula of a compound with the molecular formula N_2O_4?

(1) NO   (3) N_2O
(2) NO_2 (4) N_2O_3

37 What is the correct Lewis electron-dot structure for the compound magnesium fluoride?

(1) Mg:F:

(2) $Mg^+[:F:]^-$

(3) $[:F:]^- Mg^{2+} [:F:]^-$

(4) :F:Mg:F:

38 Given the reaction:

$Mg(s) + 2\,AgNO_3(aq) \rightarrow Mg(NO_3)_2(aq) + 2\,Ag(s)$

Which type of reaction is represented?

(1) single replacement   (3) synthesis
(2) double replacement   (4) decomposition

39 Which equation shows conservation of both mass and charge?

(1) $Cl_2 + Br^- \rightarrow Cl^- + Br_2$
(2) $Cu + 2\,Ag^+ \rightarrow Cu^{2+} + Ag$
(3) $Zn + Cr^{3+} \rightarrow Zn^{2+} + Cr$
(4) $Ni + Pb^{2+} \rightarrow Ni^{2+} + Pb$

40 The volume of a gas is 4.00 liters at 293 K and constant pressure. For the volume of the gas to become 3.00 liters, the Kelvin temperature must be equal to

(1) $\dfrac{3.00 \times 293}{4.00}$
(2) $\dfrac{4.00 \times 293}{3.00}$
(3) $\dfrac{3.00 \times 4.00}{293}$
(4) $\dfrac{293}{3.00 \times 4.00}$

41 What is the molarity of a solution containing 20 grams of NaOH in 500 milliliters of solution?

(1) 1 M
(2) 2 M
(3) 0.04 M
(4) 0.5 M

42 Which graph best represents the pressure-volume relationship for an ideal gas at constant temperature?

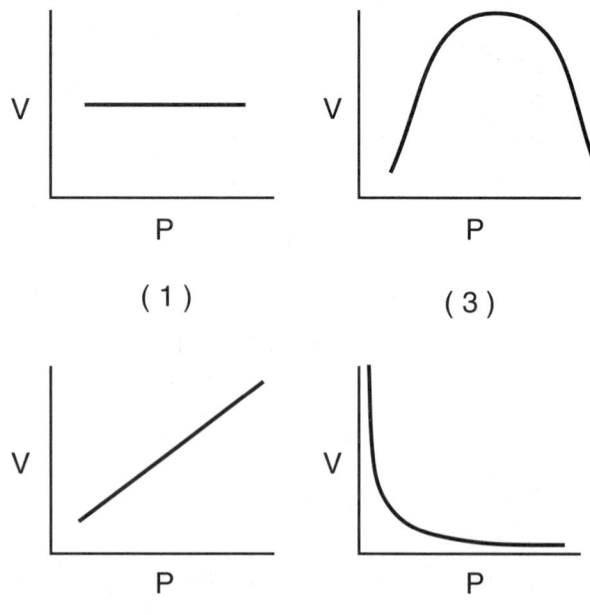

43 Given the equation:

$$C_2H_6 + Cl_2 \rightarrow C_2H_5Cl + HCl$$

This reaction is best described as

(1) addition involving a saturated hydrocarbon
(2) addition involving an unsaturated hydrocarbon
(3) substitution involving a saturated hydrocarbon
(4) substitution involving an unsaturated hydrocarbon

44 The diagram below shows a key being plated with copper in an electrolytic cell.

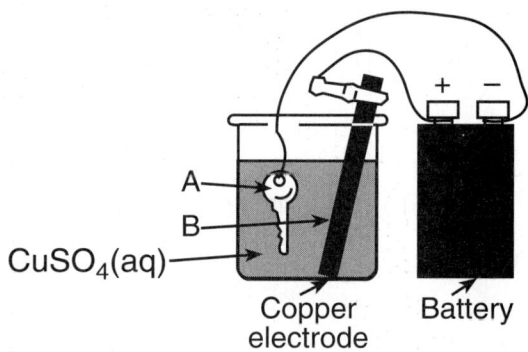

Given the reduction reaction for this cell:

$$Cu^{2+}(aq) + 2e^- \rightarrow Cu(s)$$

This reduction occurs at

(1) A, which is the anode
(2) A, which is the cathode
(3) B, which is the anode
(4) B, which is the cathode

45 A student neutralized 16.4 milliliters of HCl by adding 12.7 milliliters of 0.620 M KOH. What was the molarity of the HCl acid?

(1) 0.168 M
(2) 0.480 M
(3) 0.620 M
(4) 0.801 M

46 Nuclear fusion *differs* from nuclear fission because nuclear fusion reactions

(1) form heavier isotopes from lighter isotopes
(2) form lighter isotopes from heavier isotopes
(3) convert mass to energy
(4) convert energy to mass

47 After 32 days, 5 milligrams of an 80-milligram sample of a radioactive isotope remains unchanged. What is the half-life of this element?

(1) 8 days
(2) 2 days
(3) 16 days
(4) 4 days

48 Which electron configuration represents an atom of chlorine in an excited state?

(1) 2–8–7
(2) 2–8–8
(3) 2–8–6–1
(4) 2–8–7–1

Note that questions 49 and 50 have only three choices.

49 As each successive element in Group 15 of the Periodic Table is considered in order of increasing atomic number, the atomic radius

(1) decreases
(2) increases
(3) remains the same

50 Given the equation:

$$KNO_3(s) \xrightarrow{H_2O(\ell)} KNO_3(aq)$$

As $H_2O(\ell)$ is added to $KNO_3(s)$ to form $KNO_3(aq)$, the entropy of the system

(1) decreases
(2) increases
(3) remains the same

Part B–2

Answer all questions in this part.

Directions (51–62): Record your answers in the spaces provided in your answer booklet. Some questions may require the use of the *Reference Tables for Physical Setting/Chemistry*.

Base your answers to questions 51 and 52 on the unbalanced equation provided *in your answer booklet*.

51 Balance the equation *in your answer booklet*, using the smallest whole-number coefficients. [1]

52 *a* Using your balanced equation, show a correct numerical setup for calculating the total number of moles of $H_2O(g)$ produced when 5.0 moles of $O_2(g)$ are completely consumed. Use the space provided *in your answer booklet*. [1]

b Record your answer. [1]

Base your answers to questions 53 through 55 on the data table provided *in your answer booklet*.

53 *In your answer booklet*, record the electronegativity for the elements with atomic numbers 11 through 17. [1]

54 On the grid provided *in your answer booklet*, mark an appropriate scale on the axis labeled "Electronegativity." [1]

55 On the same grid, plot the data from the data table. Circle and connect the points. [1]

Example: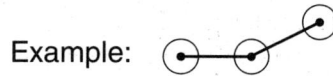

Base your answers to questions 56 through 58 on the information below.

A student uses 200 grams of water at a temperature of 60°C to prepare a saturated solution of potassium chloride, KCl.

56 Identify the solute in this solution. [1]

57 According to Reference Table G, how many grams of KCl must be used to create this saturated solution? [1]

58 This solution is cooled to 10°C and the excess KCl precipitates (settles out). The resulting solution is saturated at 10°C. How many grams of KCl precipitated out of the original solution? [1]

Base your answers to questions 59 through 61 on the diagram of the voltaic cell below.

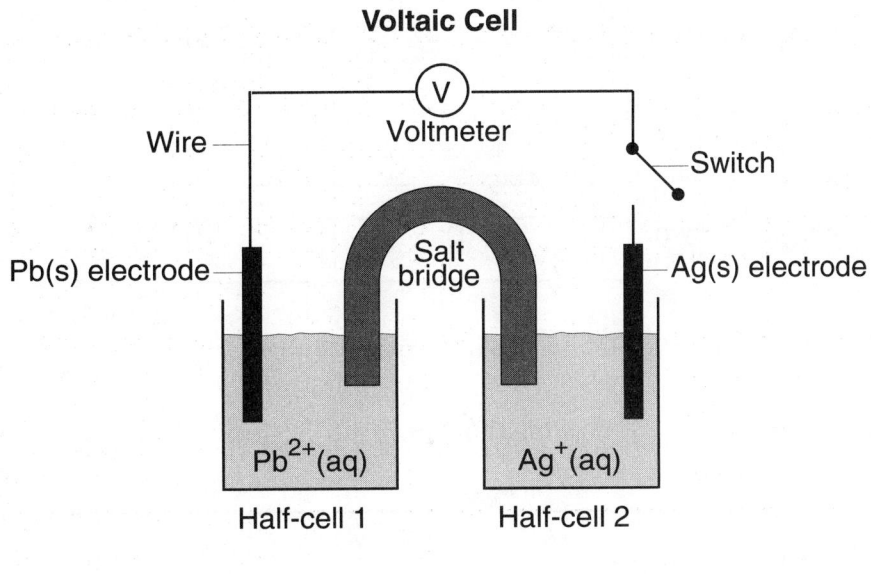

$$2\ Ag^+(aq) + Pb(s) \longrightarrow Pb^{2+}(aq) + 2\ Ag(s)$$

59 When the switch is closed, in which half-cell does oxidation occur? [1]

60 When the switch is closed, state the direction that electrons will flow through the wire. [1]

61 *Based on the given equation,* write the balanced half-reaction that occurs in half-cell 1. [1]

62 In the space provided *in your answer booklet,* draw a Lewis electron-dot structure for an atom of phosphorus. [1]

Part C

Answer all questions in this part.

Directions (63–81): Record your answers in the spaces provided in your answer booklet. Some questions may require the use of the *Reference Tables for Physical Setting/Chemistry*.

Base your answers to questions 63 and 64 on the information and the bright-line spectra represented below.

Many advertising signs depend on the production of light emissions from gas-filled glass tubes that are subjected to a high-voltage source. When light emissions are passed through a spectroscope, bright-line spectra are produced.

```
Gas A
Gas B
Gas C
Gas D
Unknown mixture
```

63 Identify the *two* gases in the unknown mixture. [2]

64 Explain the production of an emission spectrum in terms of the *energy states of an electron*. [1]

Base your answers to questions 65 through 67 on the particle diagrams below, which show atoms and/or molecules in three different samples of matter at STP.

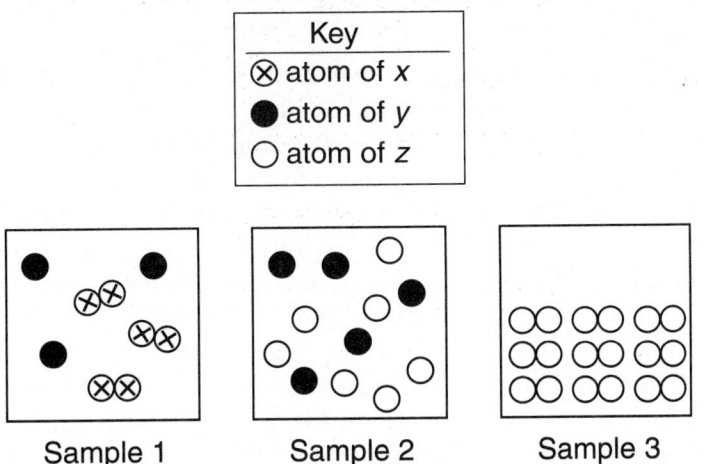

65 Which sample represents a pure substance? [1]

66 When two atoms of *y* react with one atom of *z*, a compound forms. Using the number of atoms shown in sample 2, what is the maximum number of molecules of this compound that can be formed? [1]

67 Explain why ⊗⊗ does *not* represent a compound. [1]

_____

Base your answers to questions 68 through 70 on the information below.

Many artificial flavorings are prepared using the type of organic reaction shown below.

```
     H  O                  H H H                H  O     H H H
     |  ||                 | | |                |  ||    | | |
  H- C- C -OH  +  HO- C-C-C -H   →   H- C- C -O- C-C-C -H  +  HOH
     |                    | | |                |         | | |
     H                    H H H                H         H H H

  Reactant 1            Reactant 2
```

68 What is the name of this organic reaction? [1]

69 To what class of organic compounds does reactant 2 belong? [1]

70 In the space provided *in your answer booklet*, draw the structural formula of an isomer of reactant 2. [1]

Base your answers to questions 71 through 74 on the article below, the *Reference Tables for Physical Setting/Chemistry,* and your knowledge of chemistry.

Radioactivity at home

You may be surprised to learn that you do not need to visit a nuclear power plant or a hospital X-ray laboratory to find sources of radioactivity. They are all around us. In fact, it is likely that you'll find a few at home. Your front porch may incorporate cinder blocks or granite blocks. Both contain uranium. Walk through the front door, look up, and you'll see a smoke detector that owes its effectiveness to the constant source of alpha particle emissions from Americium-241. As long as the gases remain ionized within the shielded container, electricity flows, and all is calm. When smoke enters the chamber, it neutralizes the charges on these ions. In the absence of these ions, the circuit breaks and the alarm goes off.

Indicator lights on your appliances may use Krypton-85; electric blankets, promethium-147; and fluorescent lights, thorium-229. Even the food we eat is radioactive. The more potassium-rich the food source, the more potassium-40—a radioactive isotope that makes up about 0.01% of the natural supply of this mineral—is present. Thus, brazil nuts, peanuts, bananas, potatoes, and flour, all rich in potassium, are radiation sources.

—*Chem Matters*
April 2000

71 Write the equation for the alpha decay that occurs in a smoke detector containing Americium-241 (Am-241). [2]

72 How is the radioactive decay of Krypton-85 different from the radioactive decay of Americium-241? [1]

73 State one benefit or useful application of radioactivity *not* mentioned in this article. [1]

74 State one risk or danger associated with radioactivity. [1]

Base your answers to questions 75 and 76 on the information below.

Gypsum is a mineral that is used in the construction industry to make drywall (sheetrock). The chemical formula for this hydrated compound is $CaSO_4 \cdot 2\ H_2O$. A hydrated compound contains water molecules within its crystalline structure. Gypsum contains 2 moles of water for each 1 mole of calcium sulfate.

75 What is the gram formula mass of $CaSO_4 \cdot 2\ H_2O$? [1]

76 *a* In the space provided *in your answer booklet,* show a correct numerical setup for calculating the percent composition by mass of water in this compound. [1]

  b Record your answer. [1]

Base your answers to questions 77 through 79 on the information and potential energy diagram below.

Chemical cold packs are often used to reduce swelling after an athletic injury. The diagram represents the potential energy changes when a cold pack is activated.

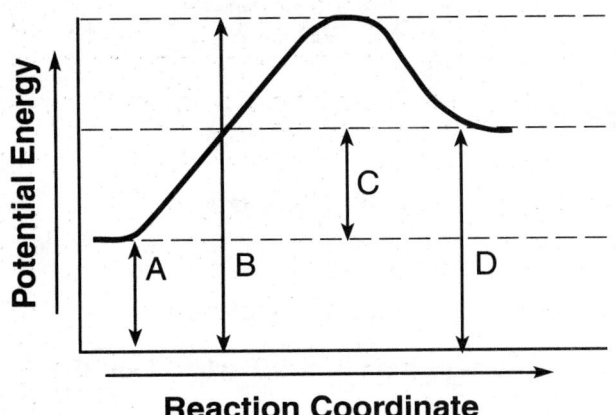

Reaction Coordinate

77 Which lettered interval on the diagram represents the potential energy of the products? [1]

78 Which lettered interval on the diagram represents the heat of reaction? [1]

79 Identify a reactant listed in Reference Table *I* that could be mixed with water for use in a chemical cold pack. [1]

Base your answers to questions 80 and 81 on the information below.

Calcium hydroxide is commonly known as agricultural lime and is used to adjust the soil pH. Before the lime was added to a field, the soil pH was 5. After the lime was added, the soil underwent a 100-fold decrease in hydronium ion concentration.

80 What is the new pH of the soil in the field? [1]

81 According to Reference Table *F*, calcium hydroxide is soluble in water. Identify another hydroxide compound that contains a Group 2 element and is also soluble in water. [1]

The University of the State of New York

REGENTS HIGH SCHOOL EXAMINATION

PHYSICAL SETTING
CHEMISTRY

Wednesday, January 28, 2004 — 9:15 a.m. to 12:15 p.m., only

ANSWER SHEET

Student .... Sex: ☐ Male ☐ Female Grade ....

Teacher .... School ....

Record your answers to Part A and Part B–1 on this answer sheet.

Part A

1	12	23 ....
2	13	24 ....
3	14	25 ....
4	15	26 ....
5	16	27 ....
6	17	28 ....
7	18	29 ....
8	19	30 ....
9	20	31 ....
10	21	**Part A Score** ☐
11	22	

Part B–1

32	42 ....
33	43 ....
34	44 ....
35	45 ....
36	46 ....
37	47 ....
38	48 ....
39	49 ....
40	50 ....
41	**Part B–1 Score** ☐

Write your answers to Part B–2 and Part C in your answer booklet.

The declaration below should be signed when you have completed the examination.

I do hereby affirm, at the close of this examination, that I had no unlawful knowledge of the questions or answers prior to the examination and that I have neither given nor received assistance in answering any of the questions during the examination.

_____
Signature

Glossary

A

accuracy a description of how close a measurement is to the true value of the quantity measured (p. 201)

acid any compound that increases the number of hydronium ions when dissolved in water; acids turn blue litmus paper red and react with bases and some metals to form salts (p. 136)

activated complex a molecule in an unstable state intermediate to the reactants and the products in the chemical reaction (p. 86)

activation energy the minimum amount of energy required to start a chemical reaction (p. 86)

addition reaction a reaction in which an atom or molecule is added to an unsaturated molecule (p. 162)

alkali metal one of the elements of Group 1 of the periodic table (lithium, sodium, potassium, rubidium, cesium, and francium) (p. 44)

alkaline-earth metal one of the elements of Group 2 of the periodic table (beryllium, magnesium, calcium, strontium, barium, and radium) (p. 44)

alkane a hydrocarbon characterized by a straight or branched carbon chain that contains only single bonds (p. 157)

alkene a hydrocarbon that contains one or more double bonds (p. 157)

alkyne a hydrocarbon that contains one or more triple bonds (p. 158)

alpha particle a positively charged atom that is released in the disintegration of radioactive elements and that consists of two protons and two neutrons (p. 186)

anion an ion that has a negative charge (p. 57)

anode the electrode on whose surface oxidation takes place; anions migrate toward the anode, and electrons leave the system from the anode (p. 176)

Arrhenius acid a substance that increases the concentration of hydronium ions in aqueous solution (p. 135)

Arrhenius base a substance that increases the concentration of hydroxide ions in aqueous solution (p. 136)

atom the smallest unit of an element that maintains the properties of that element (pp. 9, 17)

atomic mass unit a unit of mass that describes the mass of an atom or molecule; it is exactly 1/12 of the mass of a carbon atom with mass number 12 (symbol, amu) (p. 29)

atomic mass the mass of an atom expressed in atomic mass units (p. 29)

atomic number the number of protons in the nucleus of an atom; the atomic number is the same for all atoms of an element (p. 30)

average atomic mass the weighted average of the masses of all naturally occurring isotopes of an element (p. 35)

B

base any compound that increases the number of hydroxide ions when dissolved in water; bases turn red litmus paper blue and react with acids to form salts (p. 136)

beta particle a charged electron emitted during certain types of radioactive decay, such as beta decay (p. 185)

binary compound a compound composed of two different elements (p. 62)

Boyle's law the law that states that for a fixed amount of gas at a constant temperature, the volume of the gas increases as the pressure of the gas decreases and the volume of the gas decreases as the pressure of the gas increases (p. 115)

bright-line spectrum a spectrum having the pattern of a series of lines, which results from emission of electromagnetic radiation by atoms, ions, and molecules following excitations of their electrons (p. 24)

C

catalyst a substance that changes the rate of a chemical reaction without being consumed or changed significantly (p. 88)

cathode the electrode on whose surface reduction takes place (p. 177)

cation an ion that has a positive charge (p. 56)

Glossary continued

Charles's law the law that states that for a fixed amount of gas at a constant pressure, the volume of the gas increases as the temperature of the gas increases and the volume of the gas decreases as the temperature of the gas decreases (p. 116)

chemical change a change that occurs when one or more substances change into entirely new substances with different properties (p. 7)

chemical energy the energy released when a chemical compound reacts to produce new compounds (p. 85)

chemical equation a representation of a chemical reaction that uses symbols to show the relationship between the reactants and the products (p. 90)

chemical equilibrium a state of balance in which the rate of a forward reaction equals the rate of the reverse reaction and the concentrations of products and reactants remain unchanged (p. 148)

chemical property a property of matter that describes a substance's ability to participate in chemical reactions (p. 6)

chemical reaction the process by which one or more substances change to produce one or more different substances (p. 7)

combustion reaction the oxidation reaction of an organic compound, in which heat is released (p. 164)

compound a substance made up of atoms of two or more different elements joined by chemical bonds (p. 10)

concentration the amount of a particular substance in a given quantity of a mixture, solution, or ore (p. 127)

covalent bond a bond formed when atoms share one or more pairs of electrons (p. 69)

D

decomposition reaction a reaction in which a single compound breaks down to form two or more simpler substances (p. 95)

density the ratio of the mass of a substance to the volume of the substance; often expressed as grams per cubic centimeter for solids and liquids and as grams per liter for gases (p. 3)

double replacement reaction a reaction in which a gas, a solid precipitate, or a molecular compound forms from the apparent exchange of atoms or ions between two compounds (p. 95)

ductile describes the ability of a substance to be hammered thin or drawn out into a wire (p. 45)

E

electrochemical cell a system that contains two electrodes separated by an electrolyte phase (p. 175)

electrode a conductor used to establish electrical contact with a nonmetallic part of a circuit, such as an electrolyte (p. 175)

electrolysis the process in which an electric current is used to produce a chemical reaction, such as the decomposition of water (p. 179)

electrolyte a substance that dissolves in water to give a solution that conducts an electric current (p. 128)

electrolytic cell an electrochemical device in which electrolysis takes place when an electric current is in the device (p. 178)

electron a subatomic particle that has a negative charge (p. 17)

electronegativity a measure of the ability of an atom in a chemical compound to attract electrons (p. 48)

element a substance that cannot be separated or broken down into simpler substances by chemical means; all atoms of an element have the same atomic number (p. 9)

empirical formula a chemical formula that shows the composition of a compound in terms of the relative numbers and kinds of atoms in the simplest ratio (p. 108)

endothermic describes a process in which heat is absorbed from the environment (p. 3)

energy the capacity to do work (p. 3)

entropy a measure of the randomness or disorder of a system (p. 145)

enzyme a type of protein that speeds up metabolic reactions in plant and animals without being permanently changed or destroyed (p. 88)

esterification reaction an organic reaction in which an acid reacts with an alcohol to form an ester and water (p. 164)

Glossary continued

excited state a state in which an atom has more energy than it does at its ground state (p. 20)

exothermic describes a process in which a system releases heat into the environment (p. 3)

F

fermentation reaction an oxidation reaction catalyzed by enzymes in microorganisms in an environment lacking oxygen (p. 164)

formula mass the sum of the average atomic masses of all atoms represented in the formula of any molecule, formula unit, or ion (p. 105)

functional group the portion of a molecule that is active in a chemical reaction and that determines the properties of many organic compounds (p. 160)

G

gamma radiation the emission of a high-energy photon by a nucleus during fission and radioactive decay (p. 187)

gram-formula mass the mass in grams of one mole (6.02×10^{23} particles) of a substance (p. 106)

ground state the lowest energy state of a quantized system (p. 20)

group a vertical column of elements in the periodic table; elements in a group share chemical properties (p. 44)

H

half-life the time required for half of a sample of a radioactive substance to disintegrate by radioactive decay or by natural processes (p. 190)

half-reaction the part of a reaction that involves only oxidation or reduction (p. 171)

halogen one of the elements of Group 17 (fluorine, chlorine, bromine, iodine, and astatine); halogens combine with most metals to form salts (p. 44)

heat the energy transferred between objects that are at different temperatures; energy is always transferred from higher-temperature objects to lower-temperature objects until thermal equilibrium is reached (p. 4)

heterogeneous composed of dissimilar components (p. 11)

homogeneous describes something that has a uniform structure or composition throughout (p. 11)

hydrocarbon an organic compound composed only of carbon and hydrogen (p. 157)

hydrogen bond the intermolecular force occurring when a hydrogen atom that is bonded to a highly electronegative atom of one molecule is attracted to two unshared electrons of another molecule (p. 77)

hydronium ion an ion consisting of a proton combined with a molecule of water; H_3O^+ (p. 135)

I

ideal gas [no bf in unit] an imaginary gas whose particles are infinitely small and do not interact with each other (p. 118)

indicator a compound that can reversibly change color depending on conditions such as the pH of the solution or other chemical change (p. 139)

intermolecular forces the forces of attraction between molecules (p. 77)

ion an atom, radical, or molecule that has gained or lost one or more electrons and has a negative or positive charge (p. 56)

ionic bond a force that attracts electrons from one atom to another, which transforms a neutral atom into an ion (p. 58)

ionization energy the energy required to remove an electron from an atom or ion (p. 48)

isomer one of two or more compounds that have the same chemical composition but different structures (p. 158)

isotope an atom that has the same number of protons (atomic number) as other atoms of the same element do but that has a different number of neutrons (atomic mass) (p. 33)

K

kinetic-molecular theory a theory that explains that the behavior of physical systems depends on the combined actions of the molecules constituting the system (p. 118)

Glossary continued

L

law of conservation of energy the law that states that energy cannot be created or destroyed but can be changed from one form to another (p. 87)

law of conservation of mass the law that states that mass cannot be created or destroyed in ordinary chemical and physical changes (p. 91)

Le Châtelier's principle the principle that states that a system in equilibrium will oppose a change in a way that helps eliminate the change (p. 149)

Lewis structure a structural formula in which electrons are represented by dots; dot pairs or dashes between two atomic symbols represent pairs in covalent bonds (p. 71)

M

malleable describes the ability of a substance to be hammered or beaten into a sheet (p. 45)

mass number the sum of the numbers of protons and neutrons in the nucleus of an atom (p. 31)

mass a measure of the amount of matter in an object; a fundamental property of an object that is not affected by the forces that act on the object, such as the gravitational force (p. 1)

matter anything that has mass and takes up space (p. 1)

meniscus the curve at a liquid's surface by which one measures the volume of the liquid (p. 201)

metal an element that is shiny and that conducts heat and electricity well

metallic bond a bond formed by the attraction between positively charged metal ions and the electrons around them (p. 76)

metalloid elements that have properties of both metals and nonmetals; sometimes referred to as semiconductors (p. 45)

mixture a combination of two or more substances that are not chemically combined (p. 10)

molarity a concentration unit of a solution expressed as moles of solute dissolved per liter of solution (p. 127)

mole ratio the relative number of moles of the substances required to produce a given amount of product in a chemical reaction (p. 103)

mole the SI base unit used to measure the amount of a substance whose number of particles is the same as the number of atoms of carbon in exactly 12 g of carbon-12 (p. 103)

molecular formula a chemical formula that shows the number and kinds of atoms in a molecule, but not the arrangement of the atoms (p. 70)

molecule the smallest unit of a substance that keeps all of the physical and chemical properties of that substance; it can consist of one atom or two or more atoms bonded together (p. 69)

N

neutralization reaction the reaction of the ions that characterize acids (hydronium ions) and the ions that characterize bases (hydroxide ions) to form water molecules and a salt (p. 140)

neutron a subatomic particle that has no charge and that is found in the nucleus of an atom (p. 17)

noble gas an unreactive element of Group 18 of the periodic table; the noble gases are helium, neon, argon, krypton, xenon, or radon (p. 44)

nonpolar covalent bond a covalent bond in which the bonding electrons are equally attracted to both bonded atoms (p. 71)

nuclear fission the splitting of the nucleus of a large atom into two or more fragments; releases additional neutrons and energy (p. 188)

nuclear fusion the combination of the nuclei of small atoms to form a larger nucleus; releases energy (p. 189)

nucleus in physical science, an atom's central region, which is made up of protons and neutrons (p. 17)

O

orbital a region in an atom where there is a high probability of finding electrons (p. 20)

organic compound a covalently bonded compound that contains carbon, excluding carbonates and oxides (p. 157)

Glossary continued

oxidation number the number of electrons that must be added to or removed from an atom in a combined state to convert the atom into the elemental form (p. 172)

oxidation a reaction that removes one or more electrons from a substance such that the substance's valence or oxidation state increases (p. 171)

oxidation-reduction reaction any chemical change in which one species is oxidized (loses electrons) and another species is reduced (gains electrons); also called redox reaction (p. 171)

P

percent composition the percentage by mass of each element in a compound (p. 107)

period in chemistry, a horizontal row of elements in the periodic table (p. 43)

pH a value that is used to express the acidity or alkalinity (basicity) of a system; each whole number on the scale indicates a tenfold change in acidity; a pH of 7 is neutral, a pH of less than 7 is acidic, and a pH of greater than 7 is basic (p. 138)

phase equilibrium an equilibrium condition in which changes from one phase to another and back are occurring at equal rates (p. 147)

physical change a change of matter from one form to another without a change in chemical properties (p. 3)

physical property a characteristic of a substance that does not involve a chemical change, such as density, color, or hardness (p. 3)

polar covalent bond a covalent bond in which a pair of electrons shared by two atoms is held more closely by one atom (p. 71)

polyatomic ion an ion made of two or more atoms (p. 62)

polymerization reaction a reaction that forms a polymer from a series of small units (p. 163)

precision the exactness of a measurement (p. 201)

product a substance that forms in a chemical reaction (p. 7)

proton a subatomic particle that has a positive charge and that is found in the nucleus of an atom; the number of protons of the nucleus is the atomic number, which determines the identity of an element (p. 17)

Q

quantity something that has magnitude, size, or amount (p. 199)

R

radioactivity the process by which an unstable nucleus emits one or more particles or energy in the form of electromagnetic radiation (p. 185)

reactant a substance or molecule that participates in a chemical reaction (p. 7)

reduction a chemical change in which electrons are gained, either by the removal of oxygen, the addition of hydrogen, or the addition of electrons (p. 171)

S

saponification reaction a chemical reaction in which esters of fatty acids react with a strong base to produce glycerol and a fatty acid salt; the process that is used to make soap (p. 164)

saturated hydrocarbon an organic compound formed only by carbon and hydrogen linked by single bonds (p. 158)

saturated solution a solution that cannot dissolve any more solute under the given conditions (p. 127)

scientific notation a method of expressing a quantity as a number multiplied by 10 to the appropriate power (p. 204)

significant figure a prescribed decimal place that determines the amount of rounding off to be done based on the precision of the measurement (p. 202)

single replacement reaction [Text TK] [single-displacement reaction: a reaction in which one element or radical takes the place of another element or radical in a compound] (p. 95)

solubility the ability of one substance to dissolve in another at a given temperature and pressure; expressed in terms of the amount of solute that will dissolve in a given amount of solvent to produce a saturated solution (p. 123)

solute in a solution, the substance that dissolves in the solvent (p. 123)

solution a homogeneous mixture of two or more substances uniformly dispersed throughout a single phase (p. 123)

Glossary continued

solvent in a solution, the substance in which the solute dissolves (p. 123)

standard temperature and pressure for a gas, the temperature of 0°C and the pressure 1.00 atm (p. 116)

states of matter the physical forms of matter, which are solid, liquid, gas, and plasma (p. 1)

stoichiometry the proportional relationships between two or more substances during a chemical reaction (p. 103)

structural formula a formula that indicates the location of the atoms, groups, or ions relative to one another in a molecule and that indicates the number and location of chemical bonds (p. 70)

sublimation the process in which a solid changes directly into a gas (The term is sometimes also used for the reverse process.) (p. 147)

substituion reaction a reaction in which one or more atoms replace another atom or group of atoms in a molecule (p. 163)

supersaturated solution a solution that holds more dissolved solute than is required to reach equilibrium at a given temperature (p. 127)

synthesis reaction a reaction in which two or more substances combine to form a new compound (p. 95)

T

temperature a measure of how hot (or cold) something is; specifically, a measure of the average kinetic energy of the particles in an object (p. 4)

titration a method to determine the concentration of a substance in solution by adding a solution of known volume and concentration until the reaction is completed, which is usually indicated by a change in color (p. 139)

transmutation the transformation of atoms of one element into atoms of a different element as a result of a nuclear reaction (p. 185)

U

unit a quantity adopted as a standard of measurement (p. 199)

unsaturated hydrocarbon a hydrocarbon that has available valence bonds, usually from double or triple bonds with carbon (p. 158)

unsaturated solution a solution that contains less solute than a saturated solution does and that is able to dissolve additional solute (p. 127)

V

valence electron an electron that is found in the outermost shell of an atom and that determines the atom's chemical properties (p. 44)

vapor pressure the partial pressure exerted by a vapor that is in equilibrium with its liquid state at a given temperature (p. 147)

voltaic cell a primary cell that consists of two electrodes made of different metals immersed in an electrolyte; used to generate voltage (p. 178)

volume a measure of the size of a body or region in three-dimensional space (p. 1)

W

water of hydration Water present in a definite amount and attached to a compound to form a hydrate; can be removed, as by heating, without altering the composition of the compound (p. 93)

weight a measure of the gravitational force exerted on an object; its value can change with the location of the object in the universe (p. 1)

Index

Boldface page numbers refer to illustrative material, such as figures, tables, photographs, and illustrations.

A

accelerators, particle, 46, 187
accuracy, 201–202, **202**
acid-base indicators, 139, **216**
acidic solution, 138
acids
   Arrhenius, 135, 136
   defined, 136
   as electrolytes, 135
   neutralization reaction of, 140
   organic, 161, 164, **220**
   pH scale for, 138–139
   as proton donors, 136–137
   strong, 135, 139, 216
   table of common acids, **215**
   titration of, 139–140, **223**
   weak, 135
activated complex, 86
activation energy
   catalysts and, 88–89, **89**
   defined, 86
   for endothermic reaction, 87, **87**
   for exothermic reaction, 86, **86**
   reaction rate and, 88–89, **89**
activity series, 177–178, **214, 215**
addition reactions, 162–163
air, as a mixture, 10
alcohols, 160, **220**
   esterification reactions of, 164
   from fermentation, 164
aldehydes, **220**
alkali metals, 44
alkaline-earth metals, 44
alkanes, 157, **157**, 158, **219**
   isomers of, 159, **159**
   substitution reactions of, 163
alkenes, 157–158, **157, 219**
alkynes, 158, **158, 219**
allotropes, of carbon, 157
alpha particles, 186, 187, **218**
   in smoke detectors, 193
amides, **220**
amines, **220**
amount, unit of, **199**. *See also* moles
ampere, **199**
amu (atomic mass unit), 29
   formula mass and, 105–106
anions, 57
   in crystals, 64, 77
   in ionic bonds, 58–59
   names of, 59
   sizes of, 57
anode, 176, **176,** 179
aqueous solutions. *See also* solutions
   of acids and bases, 135–137, 138–140
   aq, 91, **91,** 123
   boiling points of, 128–129
   concentration of, 127–128, 139
   defined, 123
   as electrical conductors, 64, 128, 135, 175, 176
   formation of, 124
   freezing points of, 128–129
   of gases, 125–126
   pH of, 138–139, **216**
   solubility and, 123, 124–126, 127, **210, 211**
   symbol for, 91, **91,** 123
Arrhenius, Svante, 135
Arrhenius acid, 135, 136, 216
Arrhenius base, 136, 137, 216
atmosphere (atm), **207**
atomic mass, 29. *See also* elements
   average, 35, 42, **42**
   defined, 29
   in formula mass, 105
   periodic table and, 41
atomic mass unit (amu), 29
   formula mass and, 105–106
atomic number, 30–31, 32. *See also* elements
   in balanced nuclear equation, 186
   periodic table and, 41, 42, **42**
   of selected elements, **221–222**
   symbol for element and, 33–34
atomic radius, 47–48, **49**
   of selected elements, **221–222**
atoms. *See also* electrons; elements; neutrons; protons
   in a compound, 10
   defined, 9, 17
   electronegativity of, 48, **49,** 73–75, 77, **221–222**
   historical theories of, 17
   ionization energy of, 48–49, **49,** 56, **221–222**
   mass number of, 31, 32–34, 42
   models of, 18, 19–20
   parts of, 17–18, **18**
   rearranged in chemical reaction, 7
   size of, 47–48, **49, 221–222**
average atomic mass, 35
   on periodic table, 42, **42**
Avogadro, Amadeo, 117
Avogadro's law, 117

B

balancing chemical equations, 91–93, 103
balancing nuclear equations, 185–186
bases
   Arrhenius, 136, 137
   defined, 136
   as electrolytes, 135

Index continued

bases (continued)
  neutralization reaction of, 140
  pH scale for, 138–139
  as proton acceptors, 136–137
  strong, 136, 139, 216
  table of common bases, **216**
  titration of, 139–140, **223**
  weak, 136
basic solution, 138
battery, 175–177, **176**, 179
beta particles, 185, 186, 187, 190, **218**
binary compounds, 62–63
  synthesis of, 95
Bohr, Niels, 19–20
boiling points
  of aqueous solutions, 128–129
  as condensation points, 5
  of covalent compounds, 75
  of ionic compounds, 64
  of selected elements, **221–222**
  vapor pressure and, **212**
  of water, **4**, 5, 64, 78
bonds
  chemical energy of, 85
  covalent, 69–72, 74, 75
  electronegativity and, 73–75, 77
  hydrogen, 77–78, **78**
  ionic, 58–59, 63–64, 69, 70, 75
  metallic, 76–77
Boyle's law, 115
bright-line spectrum, 24
brittle, defined, 64
burning, 6, 7, 164–165

C

calculations, significant figures in, 203–204
carbon. See also organic compounds
  allotropes of, 157
  atomic mass unit based on, 29
  covalent bonds of, 69
  electron configuration of, 42
  isotopes of, 33 (see also carbon-14)
  Lewis structure of, 72
  mole of, 103
  oxidation numbers of, 173
  on periodic table, 42, **42**
carbon-14, 33
  dating specimens with, 191
  properties of, **217**
  transmutation of, 185
catalyst, 88–89, **89**
  symbol for, **91**
cathode, **176**, 177, **179**
cations, 56
  in crystals, 64, 77
  in ionic bonds, 58–59
  names of, 59
  sizes of, 57
chain reactions, nuclear, 188, 192
change, delta symbol for, 145
changes in state, 3, 4–5, **4**. See also boiling points; melting of ice; melting points
charge. See also ions
  in balanced nuclear equation, 186
  in balanced redox equation, 172
  equal to oxidation number, 173
  of ion, 56, 57
  of subatomic particles, 17, **18, 29**
Charles's law, 116
chemical change, 6–7, 85
  conservation of mass in, 91, 172
  produced by electrical current, 179
chemical energy, 4, 85
  in electrochemical cell, 176, 178
chemical equations, 7, 90–93
  balancing, 91–93, 103
  symbols in, 90–91, **91**
chemical equilibrium, 148, 149–150
chemical properties, 6, 41
chemical reactions, 6–7, 85–95
  activation energy of, 86, **86**, 87, **87**, 88–89, **89**
  addition, 162–163
  combustion, 6, 7, 164–165
  completion, 147
  decomposition, 88–89, **89**, 95, **178**, 179
  defined, 6, 85
  endothermic, 7, 85, 87, **87**, **213**
  energy changes in, 7, 85–89, 145, 146, **213**
  equations for, 7, 90–93, **91**, 103
  equilibrium in, 147–148, 149–150
  esterification, 164
  exothermic, 7, 85–87, **86**, 145, 146, 149, **213**
  fermentation, 164
  heats of, **213**
  organic, 162–165
  oxidation-reduction, 171–172, 174, 175–179
  polymerization, 163–164
  rates of, 87–89, **89**
  replacement, 95, 177–178, 215
  reversible, **91**, 147–148
  saponification, 164
  spontaneous, 145, 146, 177–178, 215
  substitution, 163
  synthesis, 95
  types of, 95
chlorine
  anions of, 57, 58, 59
  bonding of, 69, 70
  reduction of, 171–172

Index continued

coefficients, in balanced equation, 92, 93, 103
collisions, of gas particles, 118
color
    as physical property, 3
    spectrum of, 24
combustion reactions, 6, 7, 164–165
completion reactions, 147
composition, percent, 107, 223, **223**
compounds, 9, 10, **10**. *See also* covalent compounds; ionic compounds; organic compounds
    binary, 62–63, 95
    defined, 9
    formula mass of, 105–106
concentration
    determined by titration, 139
    equilibrium and, 150
    reaction rate and, 88
    of solution, 127–128, **223**
condensation of vapor, 3, 5
    at equilibrium, 147
conservation of energy, 87
conservation of mass, 91, 172, 189
conversion factors, 200
covalent bonds, 69–72
    broken in chemical reaction, 86
    electronegativity and, 74
    nonpolar, 71, 74
    polar, 71, 74, 75, 77, 173
    in structural formulas, 70, 71–72
covalent compounds
    intermolecular forces in, 77–78
    properties of, 75
crystals of ionic compounds, 64, 77
    solubility of, 124

D

Dalton (unit of mass), 29
Dalton, John, 17, 29
dating, with radioisotopes, 190–191
decimal point
    unit prefixes, 208
    in scientific notation, 204–205
    significant figures and, 203, 204
decomposition reactions, 95
    catalyzed, 88–89, **89**
    electrolytic, **178,** 179
density, 3
    calculation of, 203–204
    of selected elements, **221–222**
dissociation
    of acids and bases, 135
    of ions, 123–124, 128
double covalent bonds, 70, 72
    in alkenes, 157–158, **157, 219**
double replacement reactions, 95
ductile metals, 45

E

Einstein, Albert, 189
elastic collisions of gas particles, 118
electrical energy, 4
    in electrochemical cell, 176, 178–179
    from nuclear reactors, 192
electric current. *See also* electricity, conduction of
    electrolysis and, 179
    in electrolytic solution, 128
    unit of, **199**
electricity, conduction of. *See also* electric current
    by acid or base solutions, 135
    covalent compounds and, 75
    by electrodes, 175, 176–177
    by electrolytes, 128, 135, 175, 176
    by metals, 45, 77
    by molten salts, 64
electrochemical cells, 175–179
    activities and, 177–178, 215
    electrolytic, 178–179, **178, 179**
    voltaic, **176,** 178, 179
electrodes, 175, 176–177, 215
electrolysis, **178,** 179, **179**
electrolytes, 128, 135, 175, 176
electrolytic cells, 178–179, **178, 179**
electron configurations, 21–23
    atom stability and, 55–57
    element properties and, 47
    on periodic table, 42, **42**
electronegativity
    bond types and, 73–75, 77
    defined, 48
    hydrogen bonds and, 77
    periodic table and, 48, **49**
    table of values, **221–222**
electrons
    atomic models and, 18–20
    atomic number and, 31
    as beta particles, 185, 186, 187, 190, **218**
    in covalent bonds, 69–72
    in electrochemical cells, 175, 176, 177, 215
    electronegativity and, 48
    energy levels of, 19–20, 21, 24
    in ionic bonds, 58–59
    ionization energy and, 48–49, 56
    light emission and, 24
    mass of, **18,** 29, **29**
    oxidation numbers and, 172–173
    in oxidation-reduction reactions, 171–172, 175, 176, 177
    properties of, 17–18, **18, 29**

Index continued

elements, 9, **9.** *See also* atoms; periodic table
  activities of, 177–178, **214,** 215
  bright-line spectra of, 24
  categories of, 45–47
  in a compound, 9, 10
  defined, 9
  formula masses of, 105
  identification of, 24
  isotopes of, 32–34, 35, 42, 46–47
  mass numbers of, 31, 32–34, 42
  natural, 46
  oxidation numbers of, 173
  periodic table of, 41–49, **224–225,** 226
  properties of, **221–222**
  symbols for, 33–34, 42, **42, 221–222**
  synthetic, 46
  total number of, 31
  transmutations of, 185–187
empirical formulas, 108–109
endothermic chemical reactions, 7, 85, 87
  activation energy for, 87, **87**
  heats of reaction, **213**
endothermic physical processes, 3, 145–146
energy. *See also* heat
  activation, 86, **86,** 87, **87,** 88–89, **89**
  chemical, 4, 85, 176, 178
  chemical reactions and, 7, 85–89, 145, 146, **213**
  conservation of, 87
  covalent bond formation and, 69
  defined, 3
  of electron transitions, 24
  endothermic chemical reactions and, 7, 85, 87, **87, 213**
  endothermic physical processes and, 3, 145–146

exothermic chemical reactions and, 7, 85–87, **86,** 145, 146, 149, **213**
exothermic physical processes and, 3
forms of, 3, 4
of gas particles, 118
ionic bond formation and, 59
ionization, 48–49, **49,** 56, **221–222**
of light, 3, 24
mass converted to, 185, 189
from nuclear reactions, 185, 189, 192
physical changes and, 3–5, **4**
as product of reaction, 7, 85
solubility and, 124
total energy content of system (*H*), 145
energy levels of electrons, 19–20, 21, 24
entropy, 145–146
enzymes, 88–89, **89**
equations
  chemical, 7, 90–93, **91,** 103
  formulas for calculation of, **223**
  nuclear, 185–187, 188, 189
  word, 90
equilibrium
  in chemical reactions, 147–148, 149–150
  between phases, 147
error, percent, **223.** *See also* accuracy
esterification reactions, 164
esters, **220**
  saponification of, 164
  synthesis of, 164
estimated digit, 202–203
ethers, 160–161, **220**
evaporation, 11
  at equilibrium, 147
excited state, 20, 23, 24
exothermic chemical reactions, 7, 85–87, 145, 146

activation energy for, 86, **86,** 87
heats of reaction, **213**
Le Châtelier's principle and, 149
exothermic physical processes, 3

F

fermentation reactions, 164
filtration, 11
first ionization energy, 48, 56
fission, nuclear, 188–189, 192
fluorine, electronegativity of, 73, 74
forces
  intermolecular, 77–78, 118
  molecule-ion attraction, 124
formula mass, 105–106
formulas
  in balanced equation, 103
  for calculations, **223**
  in chemical equations, 90, 103
  empirical, 108–109
  for ionic compounds, 60–63
  molecular, 70, 108–109
  structural, 70, 71–72
freezing point, 5
  lowered by solute, 128–129
functional groups, 160–161, **220,** 221
fusion, nuclear, 189

G

gamma radiation, 187, **218**
gases. *See also* vapor
  in chemical equations, 91
  compressibility of, 115, 118
  condensation of, 3, 5, 147
  defined, 1
  elements classified as, 46
  entropy of, 146
  equilibria and, 147–148, 149–150
  ideal, 118

Index continued

kinetic energy of particles in, 115
kinetic-molecular theory of, 118
laws applying to, 115–117, 118, **223**
low density of, 115
noble, 44, 46, 55
pressure-volume relationship of, 115
solubility in liquids, 125–126
as state of matter, 1, 2, **2**
temperature of, 116–117, 118
temperature-volume relationship, 116
gas laws, 115–117, 118
   combined, **223**
glucose, fermentation of, 164
gram-formula mass, 106–107, 108
gravity, weight and, 1
ground state, 20, 22, 24
group, 44, 45
   trends in, 47, 48, 49, **49**

H

H (energy content of system), 149
   change in, 145–146, **213**
Haber process, 149–150
half-lives, of radioisotopes, 190–191
   calculations with, **223**
   table of values, **217**
half-reactions, 171–172
halogens, 44, 46
   organic compounds with, **220**
   in substitution reactions, 163
hardness, 3, 64
heat. *See also* energy
   conducted by metals, 45
   defined, 4
   for endothermic reaction, 87
   from exothermic reaction, 86, 149, **213**

formulas for, **223**
from nuclear fission, 192
reaction rate and, 88
symbol for, in equation, **91**
temperature and, 4–5, **4**
heat capacity, specific
   in formula for heat, **223**
   of water, **207**
heating curve for water, 4–5, **4**
heat of fusion, **4**, 5, 145, 146
   in heat calculation, **223**
   of water, **207**
heat of reaction, **213**
heat of vaporization, **4**, 5
   in heat calculation, **223**
   of water, **207**
heterogeneous mixtures, 11
homogeneous mixtures, 11, 123. *See also* solutions
hydration
   in ionic solids, 93
   of ions in solution, 124
hydrocarbons, 157–159, **219**
   addition reactions of, 162–163
   isomers of, 158–159, **159**
   substitution reactions of, 163
hydrogen
   atomic number of, 30–31
   in category by itself, 46
   covalent bonds of, 69, 70
   isotopes of, 46–47
   Lewis structure for, 71
   nuclear fusion of, 189
   reaction with nitrogen, 147–148, 149–150
hydrogenation, 162–163
hydrogen bonds, 77–78, **78**
hydrogen ions, 135, 136
hydrogen peroxide, decomposition of, 88–89, **89**
hydronium ions, 135, 136, 138
hydroxide ions, 136, 137, **210**

I

ideal gas, 118
identity of a substance
   change in, 6
   density and, 3
   physical change and, 3
indicators, acid-base, 139, **216**
insoluble substances, 123
   ionic compounds, 125
intermolecular forces, 77–78
   in gases, 118
ionic bonds, 58–59
   compared with covalent bonds, 69, 70
   electronegativity and, 75
   strength of, 63–64
ionic compounds. *See also* salts
   bonds in, 58–59, 63–64, 69, 70, 75
   crystals of, 64, 77, 124
   as electrolytes, 128
   forces between ions in, 77
   formulas for, 60–63
   insoluble, 125
   Lewis structures for, 72
   names of, 59, 60–61, 62
   properties of, 63–64, 75
   solubility of, 123–125, 127, **210**, **211**
ionization energy, 48–49, **49**
   cation formation and, 56
   table of values, **221–222**
ions
   in acidic or basic solutions, 135–137, 138–140
   anions, 57, 58–59, 64, 77
   cations, 56, 57, 58–59, 64, 77
   in a crystal, 64, 77, 124
   defined, 56
   dissociation of, 123–124, 128
   formation of, 55–57
   hydrated, 124
   in insoluble compounds, **210**

Index continued

ions (*continued*)
  molecule-ion attraction, 124
  names of, 59
  oxidation numbers of, 173
  sizes of, 57
  in soluble compounds, **210**
isomers, 158–159, **159**
isotopes, 32–34. *See also* radioisotopes
  average atomic mass and, 35, 42
  defined, 33
  of hydrogen, 46–47

K
ketones, **220**
kinetic energy
  defined, 4
  of particles in gas, 115
  temperature and, 4, 5
  vaporization and, 147
kinetic-molecular theory (KMT), 118

L
law of conservation of energy, 87
law of conservation of mass, 91, 172, 189
Le Châtelier, Henri, 149
Le Châtelier's principle, 149–150
length
  measurement of, 200
  unit of, **199**
Lewis structures, 71–72
  for water molecules, 77
light, 3, 4, 24
liquids. *See also* water
  in chemical equations, 91
  defined, 1
  elements classified as, 46
  entropy of, 146
  evaporation of, 11, 147
  freezing of, 3, 5, 128–129, 147

  solubility of gases in, 125–126
  as state of matter, 1, 2, **2**
  vaporization of, **4**, 5
  vapor pressure of, 147, **212**

M
malleable metals, 45
mass
  atomic, 29, 35, 41, 42, **42**, 105
  in balanced nuclear equation, 186
  conservation of, 91, 172, 189
  converted into energy, 185, 189
  converting between moles and, 106
  defined, 1
  density and, 3, 203–204
  formula mass, 105–106
  gram-formula mass, 106–107, 108
  measurement of, 200
  percentages by, 107, 223, **223**
  units of, 199, **199**
  weight and, 1
mass number
  defined, 31
  of isotopes, 32–34
  periodic table and, 42
  symbol for element and for, 33–34
matter
  changes in state of, 3, 4–5, **4**
  classifying, 9–11, **9, 10**
  defined, 1
  states of, 1–2, **2**, 3, 9
measurements
  accuracy of, 201–202, **202**
  precision of, 202, **202**
  procedures for, 200–202, **201**
  significant figures in, 202–204

  units of, 199–200, **199, 207, 208**
mechanical energy, 4
medicine, radioisotopes in, 193
melting of ice, 3, 4, **4**, 5
  entropy change, 145–146
  at equilibrium, 147
  heat of fusion and, 207
  solute effect on, 128
  temperature of, 64
melting points
  of covalent compounds, 75
  of ionic compounds, 64
  of selected elements, **221–222**
Mendeleev, Dmitri, 41
meniscus, 201, **201**, 202
metallic bonds, 76–77
metallic properties, periodic trends in, 49
metalloids, 46
metals, 45–46
  activities of, 177, **214**, 215
  alkali, 44
  alkaline-earth, 44
  electrolytic plating of, 179, **179**
  ionic bonds formed by, 63, 69
  ions of, 56
metric system, prefixes in, **208**. *See also* units
mixtures, 10–11, **10**. *See also* solutions
  heterogeneous, 11
  homogeneous, 11, 123
molarity (M), 127–128, **208, 223**
molar mass, 106. *See also* gram-formula mass
molecular formulas, 70, 108–109
molecule-ion attraction, 124
molecules. *See also* covalent compounds
  defined, 69
  forces between, 77–78, 118

Index continued

mole ratios, 103–104
moles
  converting between mass and, 106
  defined, 103
  formula for calculating, **223**
  mass of, 106–107
  percent composition and, 107
  ratios of, 103–104
  as SI units, **199**
Moseley, Henry, 41

N

names
  of ionic compounds, 59, 60–61, 62
  of ions, 59
nanotubes, 157
natural elements, 46
neutralization reaction, 140, 216
neutral solution, 138
neutrons, 17, 18
  in fission reactions, 188, 189
  mass number and, 31, 32–34, 42
  properties of, **18**, 29, **29**
  symbol for, **218**
  in transmutations, 185, 186, 187
noble gases, 44, 46, 55
nonmetals, 46
  activities of, 177, **214**, 215
  covalent bonds formed by, 69
  ionic bonds formed by, 63, 69
  ions of, 57
nonpolar covalent bonds, 71, 74
nonpolar molecules
  with polar covalent bonds, 74
  solubility of, 123
nuclear equations, 185–187, 188, 189
nuclear fission, 188–189, 192
nuclear fusion, 189

nuclear reactions, 185–193
  dating specimens with, 190–191
  electrical energy from, 192
  fission, 188–189, 192
  fusion, 189
  half-life and, 190–191, **217, 223**
  human exposure to, 192–193, **192**
  medical uses of, 193
  in smoke detectors, 193
  symbols used for, **218**
  transmutations, 185–187
  uses of, 190–193
nuclear reactors, 192
nuclear waste, 192
nucleus, 17, 18, 19, 20, 21. *See also* neutrons; protons
  atomic number and, 30–31, 32, 186
  mass number and, 31, 32–34
  unstable, 185–187, 188–189

O

orbitals, 20, 21
organic compounds, 157–165
  defined, 157
  functional groups of, 160–161, **220**, 221
  hydrocarbons, 157–158, 162–163, **219**
  isomers of, 158–159, **159**
  prefixes in names of, **218**
  reactions of, 162–165
oxidation
  activity series and, 177–178, 215
  at anodes, 176
  defined, 171
oxidation numbers, 172–174
  defined, 172
  formulas for ionic compounds and, 61–62
  on periodic table, 61–62, 173

  single-replacement reaction and, 178
oxidation-reduction reactions, 171–172
  defined, 171
  electrochemical cells and, 175–179
  oxidation numbers and, 174
oxygen
  in air, 10
  bonding of, 70, 71
  combustion and, 6, 7, 164–165
  as an element, 9, 10
  nucleus of, 31
  oxidation numbers of, 173
  properties of, 6
ozone, 9

P

particle accelerators, 46, 187
particles. *See also* atoms; molecules; subatomic particles
  of compounds, 10, **10**
  of elements, 9, **9**
  of gases, 1–2, **2**, 115, 117, 118
  of mixtures, 10, **10**
  states of matter and, 1–2, **2**
  thermal energy and, 4, 5
percent composition, 107, 223, **223**
percent error, **223**
periodic table, 41–49, **224–225, 226**. *See also* elements
  categories of elements in, 45–47
  groups, 44, 45, 47, 48, 49, **49**
  history of, 41
  information contained on, 41–42
  oxidation states on, 61–62, 173
  periods, 43–44, 45, 47, 48, 49, **49**
  trends in, 47–49, **49**
periods, 43–44, 45
  trends in, 47, 48, 49, **49**

Index continued

pH, 138–139
  indicators of, 139, **216**
phase equilibrium, 147
physical changes
  changes in state, 3, 4–5, **4**
  conservation of mass in, 91
  defined, 3
  entropy and, 145–146
physical properties
  of components in mixtures, 11
  defined, 3
  trends in periodic table, 41, 47–49
plating, electrolytic, 179, **179**
polar covalent bonds, 71, 74, 75
  oxidation numbers and, 173
  in water molecules, 77
polar molecules
  hydrogen bonds of, 77–78
  intermolecular forces in, 77, 78
  solubility and, 123, 124
polyatomic ions, 62, 63. *See also* ionic compounds; ions
  balancing equations containing, 93
  covalent bonding in, 75
  formula mass of compounds containing, 106
  solubility in water of, 124–125
  table of, **209**
polymerization reactions, 163–164
polymers, 164
positrons, 187, **218**
potential energy content, 145
precision, 202, **202**
pressure. *See also* gases
  equilibrium and, 149–150
  gas solubility and, 126
  gas volume and, 115
  ideal-gas behavior and, 118
  standard, 116, **207**
  units of, 116, **207**
  vapor pressure, 147, **212**

products. *See also* chemical reactions
  in chemical equations, 7, 90–91, **91**, 103
  defined, 7
  in endothermic processes, 87
  energy as, 7, 85
  energy content of, 145
  in exothermic processes, 86
  in nuclear equations, 186
protons, 17, 18, 19, 21
  in acid or base solutions, 136–137
  atomic number and, 30–31, 32
  mass number and, 31
  properties of, **18**, 29, **29**
  symbol for, **218**
  in transmutations, 185, 186
pure substances, 9. *See also* compounds; elements

Q

quantity, 199

R

R, in formula for organic compound, 161, **220**
radiation
  gamma rays, 187, **218**
  human exposure to, 192–193, **192**
radioactivity, 185–187, 188–189
  decay formulas, **223**
  half-life and, 190–191, **217, 223**
radioisotopes
  carbon-14, 33, 185, 191, **217**
  decay modes of, 190
  half-lives of, 190–191, **217, 223**
  human exposure to, 192–193, **192**
  medical uses of, 193
  in nuclear reactors, 192
  table of selected radioisotopes, **217**

radius
  atomic, 47–48, **49**, **221–222**
  ionic, 57
randomness, 145
rate of reaction, 87–89, **89**
reactants. *See also* chemical reactions
  in chemical equations, 7, 90–91, **91**
  concentration of, 88
  defined, 7
  energy as, 85
  energy content of, 145
  in nuclear equations, 186
reaction rate, 87–89, **89**
reactivity, electron configurations and, 55–57
redox reactions, 171–172
  defined, 171
  electrochemical, 175–179
  oxidation numbers and, 174
reduction
  at cathodes, 177
  defined, 171
rem, 192, **192**, 193
replacement reactions, 95
  activity series and, 177–178, 215
reversible reactions, **91**, 147–148
ring compounds, 159, **159**

S

salts, 59. *See also* ionic compounds; sodium chloride
  crystals of, 64, 77, 124
  electricity conducted by, 64
  formulas of, 60–61
  from neutralization reactions, 140, 216
  water of hydration in, 93
saponification reactions, 164
saturated hydrocarbons, 158. *See also* alkanes
saturated solution, 127, **211**
scientific notation, 204–205

Index continued

shells, electron, 20, 21–23, 24. *See also* electrons; valence electrons
  atomic radius and, 48
  occupied, 43–44, 45
  outermost, 44
  periodic table and, 42, 43–44
significant figures, 202–204
single covalent bond, 70
single replacement reaction, 95
  activity series and, 177–178, 215
SI units, 199–200, **199**, 209
smoke detectors, 193
sodium, oxidation of, 171–172
sodium chloride, 58–59, 61, 62, 64
  as electrolyte, 128
  freezing point and, 128
  molarity of solution of, 127
  oxidation-reduction and, 171–172
  solubility in water, 123–124, 125
solids
  in chemical equations, 91
  defined, 1
  elements classified as, 46
  entropy of, 146
  ionic, 64
  as state of matter, 1, 2, **2**
  sublimation of, 147
solubility, 123, 124–126
  defined, 123
  of gases in liquids, 125–126
  guidelines for, **210**
  of ionic compounds in water, 123–125
  "like dissolves like," 123
  temperature and, 125, 127, **211**
solubility curves, 125, 127, **211**
solute, defined, 123. *See also* solutions

solutions, 123–129. *See also* aqueous solutions
  of acids and bases, 135–137, 138–140
  boiling points of, 128–129
  components of, 123
  concentration of, 127–128, **223**
  defined, 123
  electrical conduction by, 128, 135, 175, 176
  formation of, 124
  freezing points of, 128–129
  of gases in liquids, 125–126
  pH of, 138–139, **216**
  saturated, 127, **211**
  solubility and, 123, 124–126, 127, **210, 211**
  supersaturated, 127, **211**
  unsaturated, 127, **211**
solvent, defined, 123. *See also* solutions
specific heat capacity
  in formula for heat, **223**
  of water, **207**
spectrum, bright-line, 24
spontaneous chemical reactions, 145, 146
  activity series and, 177–178, 215
spontaneous nuclear reactions, 185–187, 188
  half-lives of, 190–191, **217, 223**
stability of atoms
  covalent bonds and, 69, 70, 71
  ion formation and, 55–57
standard temperature and pressure (STP), 116, **207**
states of matter, 1–2, **2**, 9. *See also* gases; liquids; solids
  changes in, 3, 4–5, **4**
stoichiometry, 103–109
  defined, 103
  empirical formulas in, 108–109

  formula mass in, 105–106
  gram-formula mass in, 106–107, 108
  molecular formulas in, 108–109
  moles in, 103–104, 106–107, **223**
  percent composition in, 107
STP (standard temperature and pressure), 116, **207**
strong acids, 135, 139, 216
strong bases, 136, 139, 216
strong electrolytes, 135
structural formulas, 70
  of isomers, 158–159, **159**
  Lewis structures, 71–72
subatomic particles, 17–18, **18, 29**. *See also* electrons; neutrons; protons
  in nuclear chemistry, 185–187, **218**
sublimation, 147
substitution reactions, 163
sugars
  aqueous solutions of, 123, 128
  fermentation of, 164
supersaturated solution, 127, **211**
surface area, reaction rate and, 88
synthesis reactions, 95
synthetic elements, 46

T

temperature
  Celsius, 116, **207**
  changes in state and, 4–5, **4**
  energy and, 4–5, **4**
  equilibrium and, 147, 149
  of gas, 116–117, 118
  heat and, 4–5, **4**
  kelvin (K), 116, **199, 207**
  measurement of, 202–203
  reaction rate and, 88
  solubility and, 125, 127, **211**

Index continued

temperature (*continued*)
    standard, 116, **207**
    units of, 116, **199, 207**
    vapor pressure and, 147, **212**
thermal energy, 4. *See also* heat
thermodynamic temperature, **199**
thermometer, 202–203
titration, 139–140, **223**
transmutations
    artificial, 187
    spontaneous, 185–187
triple covalent bonds, 70, 72
    in alkynes, 158, **158, 219**

U

units, 199–200, **199**
    conversion factors for, 200
    prefixes for, **208**
    for pressure, 116, **207**
    table of selected units, **208**
unsaturated hydrocarbons, 158, **219**
    additions to, 162–163
unsaturated solution, 127, **211**
unstable nuclei, 185–187, 188–189

V

valence electrons. *See also* electrons; shells, electron
    in covalent bonds, 69, 70, 71–72
    defined, 44
    electronegativity and, 48
    ion formation and, 55–57
    ionization energy and, 48
    in Lewis structures, 71–72
    in metallic bonds, 76–77
    periodic table and, 44, 45
vapor. *See also* gases
    change of liquid to, **4,** 5
    condensation of, 3, 5, 147
    particles of, 2, **2,** 5
vaporization, heat of, **4,** 5
    in heat calculation, **223**
    for water, **207**
vapor pressure, 147, **212**
voltaic cell, **176,** 178, 179
volume
    defined, 1
    density and, 3, 203–204
    of gas, 1, 115, 116, 117
    gas pressure and, 115, 117
    gas temperature and, 116, 117
    of liquid, 1
    measurement of, 200–202, **201**
    of solids, 1
    units of, 199–200

W

water. *See also* aqueous solutions
    boiling of, 3, **4,** 5, 64, 78
    as a compound, 9, 10
    electrolysis of, 178–179, **178**
    freezing of, 3, 5, 128–129, 147
    heating curve, 4–5, **4**
    hydrogen bonds in, 77–78, **78**
    Lewis structure for, 77
    in nuclear reactors, 192
    physical constants for, **207**
    as polar molecule, 77–78, 123, 124
    reaction with alkali metals, 44
    states of, 2, **2**
    vaporization of, **4,** 5, **207**
water of hydration, 93
weak acids, 136
weak bases, 136
weak electrolytes, 135
weight, 1. *See also* mass
word equations, 90
work, 3

Z

zeros, significant figures and, 203